TAKE BACK HIGHER EDUCATION

Race, Youth, and the Crisis of Democracy in the Post–Civil Rights Era

HENRY A. GIROUX
and
SUSAN SEARLS GIROUX

TAKE BACK HIGHER EDUCATION
© Henry A. Giroux and Susan Searls Giroux, 2004.

First published in hardcover in 2004 by Palgrave Macmillan
First PALGRAVE MACMILLAN™ paperback edition: April 2006
175 Fifth Avenue, New York, N.Y. 10010 and
Houndmills, Basingstoke, Hampshire, England RG21 6XS.
Companies and representatives throughout the world.

PALGRAVE MACMILLAN is the global academic imprint of the Palgrave Macmillan division of St. Martin's Press, LLC and of Palgrave Macmillan Ltd. Macmillan® is a registered trademark in the United States, United Kingdom and other countries. Palgrave is a registered trademark in the European Union and other countries.

ISBN 1–4039–7290–7

Library of Congress Cataloging-in-Publication Data

Giroux, Henry A.
  Take back higher education: race, youth, and the crisis of
  democracy in the post–Civil Rights Era / by Henry A. Giroux and
  Susan Searls Giroux.
      p. cm.
  Includes bibliographical references and index.
  ISBN 1–4039–6423–8 hardback
  ISBN 1–4039–7290–7 paperback
    1. Education, Higher—Aims and objectives—United States.
  2. Universities and colleges—United States—Sociological aspects.
  3. Higher education and state—United States. 4. Politics and
  education—United States. I. Giroux, Susan Searls, 1968– II. Title.

LA227.4.G57 2004
378.73—dc22   2003066428

A catalogue record for this book is available from the British Library.

Design by Newgen Imaging Systems (P) Ltd., Chennai, India.

First edition: April 2006
10 9 8 7 6 5 4 3 2 1

Printed in the United States of America.

Additional Praise for *Take Back Higher Education*:

"Henry and Susan Giroux are performing an immense public service with this book. It is a sweeping critique of how our culture, especially the educational establishment, has failed to prepare us for the crises of our time. And it offers hope for the possibility of resisting that and creating a new culture, inspirational and profoundly democratic."

—Howard Zinn, author of *A People's History of the United States*

"Henry and Susan Giroux's extraordinary book is an electrifying call to educators to renounce political passivity and to assume the role of public intellectuals prepared to take back schools and universities from the predations of a business driven ideology that silences dissent and undermines democracy. A beautifully fashioned work of cultural history, it is also a rich and stimulating brew of elegant analysis and powerful polemic. Teachers from the kindergarten classrooms to the ivory tower will be grateful for the hope and affirmation the Girouxs' have given us. All in all, a magnificent achievement."

—Jonathan Kozol

"Here, at last, is a critical study in the social sciences that explores with brilliant iconoclasm the connection between the post 9/11 dedemocratization of America, the erosion of its politics and its civil rights, its inexorable drift into rabid conservatism, and recent attacks on the form and substance of higher education. Argued with enormous conviction and considerable insight, *Take Back Higher Education* does for contemporary pedagogy what the likes of John Dewey did for it long ago: insist that the health of our society depends not on consumption or the rampant production of wealth for the rich, but on educating new generations of citizens for open, informed public engagement, for constructive political involvement, for commitment to a social world built on justice and empowerment for all; in short, for all the things currently under threat in the security-obsessed, frightened USA of the early twenty-first century. Here, in other words, is a charter for real freedom through enlightenment, a charter that ought to have been accomplished two hundred or so years ago, but still requires a good fight. Henry and Susan Giroux have undertaken that fight with vigor, energy, and consummate intelligence."

—John Comaroff, University of Chicago

"*Take Back Higher Education* is a big book—broad, synthetic, passionate, unsparing in its analysis of the devastating implications for democracy and politics of an increasingly authoritative corporate neo-liberal state. While grounding their argument in the insights of cultural studies, critical theory and intellectual history, the Girouxs are as wary of the insularity of specialized academic languages as they are of spectacle and media culture, and of the privatized ideal of a separate peace. They leave no doubt that education is the pivotal arena for the progressive struggle for equity, agency, imagination and hope, and that this struggle must now be waged publicly, on multiple fronts, with unflinching courage and purpose."

—Michael Brenson, Avery Graduate School of the Arts,
Bard College

"Henry and Susan Giroux have written an extraordinary book explaining why we must fight to see that higher education serves democratic values. It is immensely readable. Drawing from a rich and diverse array of sources, classical and contemporary, the authors make a compelling case that the battle for the soul of higher education is at the center of the battle for the future of this nation."

—Robert McChesney, Department of Communications,
University of Illinois at Urbana-Champaign,
author of *Our Media, Not Theirs*

"This is an important volume. Based on careful analyses of American higher education's leading institutional and intellectual developments, the authors show how academia is implicated in the wider society's retreat from democracy, racial inclusion, and social justice. Giroux and Giroux make a passionate, richly informed case for the university's role as a safe and democratic place, not a haven for the privileged and powerful few. *Take Back Higher Education* belongs on the shelves of all concerned public thinkers, including the all-too-rare intellectuals who survive within the interstices of the corporate, neo-liberal university."

—Dr. Paul Street, Vice President for Research and Planning,
Chicago Urban League

*For Brett, Chris, and Jack Giroux and young people
everywhere in their struggle for knowledge, dignity,
and justice*

Contents

Acknowledgments

Books are never just the results of writers toiling alone in front of computers, hidden in out-of-the-way rooms, cubicles, or offices. They grow out of discussions with students, friends, and colleagues. They also build upon traditions of scholarship, papers heard at academic conferences, the insights of visiting lecturers, and the rush of events that shape everyday life and are analyzed daily in the dominant and independent media. For both of us, one of the most significant reasons for writing this book came from the experiences we have had over the last few decades as academics in four universities. We have been immensely influenced by the energy, convictions, and compassion of our students, as well as a number of courageous colleagues. This book is dedicated to their moral energy, critical inquisitiveness, and the hope they inspire in dire times.

Susan would like to acknowledge her gratitude to the Spencer Foundation for its generous support of her work. A fellowship enabled the completion of her dissertation, from which sections of chapters 1 and 4 are taken. This book would not have been written without the help of many friends who offered invaluable criticisms and support. We would like to especially thank Arif Dirlik, Imre Szeman, John and Jean Comaroff, Michael Brenson, Norman Denzin, Nick Couldry, Mike Payne, Robert Ivie, Robert Yarber, Sarah Schwartz, Paul Street, Jackie Edmundson, Stanley Aronowitz, Evan Watkins, Jeff Nealon, Doug Kellner, Roger Simon, Ken Saltman, Lynn Worsham, David Theo Goldberg, Lawrence Grossberg, Sut Jhally, Donaldo Macedo, Robin D. G. Kelley, Peter Trifonas, Ted Striphas, Micaela Amato, Vorris Nunley, Don Schule, Dean Birkenkamp, Jane Gordon, Lewis Gordon, Pat Shannon, and

David Gamson. Needless to say, we bear the responsibility for the final outcome. We also want to thank Brett, Chris, and Jack for reminding us why we wrote this book. Jessica Marlow, mother and mother-in-law, has never failed with her deeply appreciated love, generosity, and support. We would also like to thank Linda Barberry and Virginia Monet. As usual, our editors, Michael Flamini and Amanda Johnson, provided support right from the start of the project. Both are great editors. Henry also wants to thank Sue Stewart, his assistant, for the great sense of humor, administrative help, and overall support. We are deeply indebted to Christopher Robbins, who provided detailed editorial suggestions on the entire manuscript. Once again, we are appreciative that Grizz, our aging canine companion, is still around. We also want to thank our students, who have been of enormous help in enabling us to think through many of our arguments. While we actively collaborated on all of these chapters, there was a division of labor around the actual writing. Susan Searls Giroux wrote chapters 1, 4, and 5, and Henry Giroux wrote chapters 2, 3, and 6. We coauthored the introduction and chapter 7. Portions of some of Henry Giroux's chapters draw from ideas published previously in *Social Identities, Communication and Critical/Cultural Studies, Harvard Educational Review*, and *JAC*. Earlier versions of Susan Searls Giroux's chapters were published in *JAC* and *Social Identities*. A version of the introduction was published in *Tikkun*.

Introduction: Why Taking Back Higher Education Matters

This book represents our desire to offer some speculation about and critical questioning of two events in post–civil rights America. The first has to do with the current state of American political culture: the declining interest in and cynicism about mainstream national politics, its decidedly negative impact on the democratic process, and how such entrenched dispositions might be reversed. Emptied of any substantial content, democracy appears imperiled because individuals are unable to translate their privately suffered misery into broadly shared public concerns and collective action. Civic engagement now appears impotent and public values have become expendable as a result of the growing power of multinational corporations to shape the content of most mainstream media. Political exhaustion and impoverished intellectual visions are fed by the increasingly popular assumption that there are no alternatives to the present state of affairs.[1] For many people today, citizenship is about the act of buying and selling commodities (including political candidates), rather than broadening the scope of their freedoms and rights in order to expand the operations of a substantive democracy. Market values, coupled with a resurgent bigotry, undercut the possibility of a language in which vital social institutions can be defended as a public good. And as social visions of equity recede from public memory, unfettered brutal self-interests combine with retrograde social policies to make security and safety a top domestic priority. One consequence is that all levels of government unrelated to the military and police are being hollowed out, reducing their role to dismantling the gains of the welfare state. Increasingly, they construct policies that now criminalize social problems and prioritize penal methods over social investments. The public realm ceases to resonate as a site of democratic possibilities, as a fundamental space for how we reactivate our political sensibilities and conceive of ourselves as critical citizens, engaged public intellectuals, and social agents. The growing punitiveness and injustice in American society is in proportion to widening inequality and lack of political imagination and collective hope.[2]

The second development has to do with the nation's increasing skepticism (even overt hostility) toward the educational system at all levels. Distrust of politicians has become something of a national pastime, but why the equal disdain toward educators? Equal opportunity to attend quality educational institutions—both K-12 schools and higher education—was one of the defining principles of the Civil Rights movement, and it proved to be a focus of that era's most potent victories, as schools and universities became more accessible to minorities, women, and students with disabilities. But the backlash was immediate, and discontent with programs like affirmative action and busing was quite visible by the mid-1970s among white working-class and middle-class voters, who were feeling pinched by recession and ignored by the federal government. Conservatives eager to reach out to this constituency (and unwilling to constrain big business in the interests of working families) adopted a strategy to address their educational if not their economic concerns. Yet, they were careful to avoid the overt racism of their predecessors who attacked the rights of minority children to attend desegregated schools and to have access to post-secondary education. Both public and higher education now came under attack, though in different ways. Public schooling was increasingly redefined as a private rather than a public good. And with the shift away from public to private interests, privatization and choice became the catch phrases dominating educational reform for the next few decades. The attack on all things public was accompanied by attempts to empty the public treasury; public education became one of the first targets of neoliberals, neoconservatives, and fundamentalists advocating market interests over social needs and democratic values. With the publication of *A Nation at Risk*, the

Reagan administration gave the green light to pass spending cuts in education—cuts that have been obligatory for each administration to follow. Reconceived as a "big government monopoly," public schooling was derided as bureaucratic, inefficient, and ineffectual, producing a product (dimwitted students) who were singularly incapable of competing in the global marketplace. In short, schools had committed "an unthinking unilateral educational disarmament," the report accused. A clever strategy to be sure, which provided a ready scapegoat to legitimate the flight of U.S. manufacturing to markets overseas. Schools were blamed for increased joblessness and insecurity—not the rapacious greed of corporations eager to circumvent U.S. minimum wage laws, federal taxes, and environmental regulations, while breaking the back of unions at home. Ironically, two-plus decades of conservative railing against and radical defunding of public schools has paved the way for the offshore outsourcing of high-paying white collar jobs—the hot new thing in corporate America. In a recent interview with *The San Jose Mercury News*, the chief executive of Intel, Craig Barrett, talked about the integration of India, China, and Russia into the new global economy in the following terms: "I don't think this has been fully understood in the United States. If you look at India, China, and Russia, *they all have strong educational heritages*. . . . The big change today from what's happened over the last 30 years is that *it's no longer just low-cost labor that you are looking at. It's well educated labor* that can effectively do any job that can be done in the United States."[3] Similarly, higher education was accused of harboring a hotbed of leftist academics promoting culture wars that derided Western civilization (and in the post–9/11 era, dissent was to be equated with treason). Higher education was portrayed as the center of a class and race war in which the values and dreams of the white working class were under attack because of the ideological residue of liberal professors tainted by the legacy of radical sixties politics. The division and distrust between "elitist liberals" and a white working class was now complete and utterly sedimented. Reinventing Nietzsche's "mobile army of metaphors," the right redeployed Cold War rhetoric against the nation's schools and succeeded in a propaganda campaign to turn the popular tide against public and higher education.

Though there is nothing new in pointing out these two tragic developments—the popular retreat from politics and disdain of education—we are struck by the fact that they are rarely dealt with together as mutually reinforcing tendencies. Yet, an educated and active citizenry is indispensable for a free and inclusive democratic society; democratic politics requires the full participation of an informed populace. It requires a public willing to question and challenge its elected officials and its laws—and change both when necessary. One can't happen without the other; and now it appears both are in jeopardy.

As we've indicated, neither the decline of democracy nor the crisis of education have gone unnoticed. But curiously, the progressive advocates and activists clustered around either issue have little regard for each other. Astute readers of the national political scene have little interest in (and are often woefully ignorant of) the state of education beyond a heartfelt sound bite or two. And educators seem to have lost the language for linking schooling to democracy, convinced that education is now about job training and competitive market advantage. Yet both sides hold common concerns and seek deliberative action to change public opinion. Both suffer, however, from the retreat from what were core American values—a concern for notions of publicness, equal access and opportunity, equality and autonomy. If the liberal left seems particularly impotent and disheveled at this point in history, conservatives appear to be the masters of persuasion and organization. Working for decades at grass-roots organizing, they have taken both pedagogy and politics deadly seriously. Conversely, mainstream Democrats make no mention of an educational agenda that differs significantly from the one adopted by the Bush administration. Indeed they appear split on the far more general issue of a national platform, the majority wanting to occupy a kinder, gentler republicanism, while a few "radicals" seek to reclaim the liberal traditions of the Democratic party. Yet we argue that education—both formal and informal, public and higher—should be their first priority.

As a clear example of what we mean, consider the following statistics: soon after the invasion of Iraq, the *New York Times* released a survey indicating that 42 percent of the American public believed that Saddam Hussein was directly responsible for the

September 11 attacks on the World Trade Center and the Pentagon. CBS also released a news poll indicating that 55 percent of the public believed that Saddam Hussein directly supported the terrorist organization Al Qaeda. A majority of Americans also believed already that Saddam Hussein had weapons of mass destruction, that such weapons had been found, that he was about to build a nuclear bomb, and that he would unleash it eventually on an unsuspecting American public. A Knight Ridder/Princeton Research poll found that "44% of respondents said they thought 'most' or 'some' of the September 11, 2001, highjackers were Iraqi citizens." A poll conducted by *The Washington Post* near the second anniversary of the September 11 tragedy indicated that 70 percent of Americans continued to believe that Iraq played a direct role in the planning of the attacks. None of these claims had any basis in fact, as no evidence existed even remotely to confirm these assertions. What does this represent, if not a crisis of pedagogy—both formally and informally—in the public sphere?

Of course, these opinions did not simply fall from the sky; they were ardently legitimated by President Bush, Vice President Cheney, Colin Powell, and Condolezza Rice, and reproduced daily by an uncritical lapdog media. These misrepresentations and strategic distortions circulated in the popular press either with uncritical, jingoistic enthusiasm, as in the case of the Fox News Channel, or through the dominant media's refusal to challenge such claims— both positions, of course, in opposition to foreign news sources such as the BBC, which repeatedly challenged such assertions. Such deceptions—as the claim that Iraq was stockpiling biological and chemical weapons—are never innocent, and in this case appear to have been shamelessly used by the Bush administration to both muster support for the Iraqi invasion and for an ideological agenda "that overwhelmingly favors the President's wealthy supporters and is driving the federal government toward a long-term fiscal catastrophe."[4]

The conservative assault on public education has only strengthened under the Bush administration. As Senator Robert Byrd stated in a Senate floor speech, President Bush has no trouble asking congress for $87 billion in supplemental funds to rebuild Iraq, but refuses to allocate the $6 billion needed to fund his educational reform program, or for that matter the resources needed to maintain educational programs for our neediest students. As Byrd put it,

> I wonder how the Senators who object to the cost of my amendment . . . to add $6.1 billion for Title 1 education programs to fully fund money Congress authorized for fiscal year 2004 . . . will view the President's request to add $60 billion or $65 billion or $70 billion to the deficit to fund military and reconstruction activities in Iraq. I wonder if they will be comfortable voting to support a massive spending program for Iraq if they cannot bring themselves to support a comparatively meager increase in education funding for American schoolchildren.[5]

Of course, it will not only be schoolchildren who will be suffering from budget shortfalls, but also university students who have to grapple with skyrocketing tuition, decreasing student aid, fewer course offerings, which delay student graduation, and a generally watered-down education.

While not downplaying the seriousness of government deception, we believe there is another crucial issue that underlies these events in which the most important casualty is not simply the integrity of the Bush administration, but democracy itself. One of the central legacies of modern democracy, with its roots in classical republicanist liberal traditions, and most evident in the twentieth century in the work of W. E. B. Du Bois, Bertrand Russell, Jane Addams, and John Dewey, among others, is the important recognition that a substantive democracy simply cannot exist without educated citizens. Of course, these views have not been held universally. For some, the fear of democracy itself translated into an attack on a truly public and accessible education for all citizens. For others, such as Walter Lippman, who wrote extensively on democracy in the 1920s, representative democracy entailed creating two modes of education—one mode would be for the elite, who would rule the country and be the true participants in the democratic process, and the other branch of education would be designed for the masses, whose education would train them to be obedient workers and passive spectators rather than participants in shaping democratic public life. Progressives like Du Bois, Dewey, and

Addams however rejected such a bifurcation of educational opportunity outright.[6] They believed that education for a democratic citizenry was an essential condition of equality and social justice and had to be provided through public and higher education.

Although Dewey, Du Bois, and others were correct in linking education and democracy, they had no way in their era of recognizing that the media culture would extend, if not supercede, institutionalized education as the most important educational force in developed societies. In fact, education and pedagogy were synonymous with schooling in the public mind. Challenging such a recognition does not invalidate the enormous importance of formal education to democracy, but it does require a recognition of how the work of education now takes place in a range of spheres, including the news, advertising, television, film, the Internet, video games, and the popular press. Yet the dispersion of education only underscores with renewed urgency the significance of formal spheres of learning. Unlike their commercially driven popular counterparts, formal sites of pedagogy must provide citizens with the kinds of critical capacities, modes of literacies, knowledge, and skills that enable them to both read the world critically and participate in shaping and governing it. We are not claiming that higher education is a disinterested space, but that in its best moments it works through altogether different interests than the commercial values promoted by corporations. Instead, it self-consciously educates future citizens capable of participating in and reproducing a democratic society. In spite of their present embattled status and contradictory roles, universities and colleges remain uniquely placed to prepare students to both understand and influence the larger educational forces that shape their lives. By virtue of their privileged position and dedication to freedom and democracy such institutions also have an obligation to draw upon those traditions and resources capable of providing a liberal and humanistic education to all students in order to prepare them for a world in which information and power have taken on new and powerful dimensions.[7]

This book differs from most treatments of higher education. In the first section, it argues that higher education cannot be separated from the imperatives of an inclusive democracy and that the crisis of higher education must be understood as part of the wider crisis of politics, power, and culture. Recognizing the inextricable link between education and politics is central to reclaiming higher education as a democratic public sphere. Equally important is the recognition that politics cannot be separated from pedagogy and the sphere of culture. Of course, acknowledging that pedagogy is political because it is always tangled up with power, ideologies, and the acquisition of knowledge and skills necessary for critical participation in public life does not mean that it is by default propagandistic, closed, dogmatic, or uncritical of its own authority. Most importantly, any viable notion of critical pedagogy must demonstrate that there is a difference between critical pedagogical practices and propagandizing, critical teaching and demagoguery. Such a pedagogy should be open and discerning, fused with a spirit of inquiry that fosters rather than mandates critical modes of individual and social agency. In this context, pedagogy should not be viewed as an a priori method or a set of teaching skills, but an object of struggle over assigned meanings, modes of expression, and directions of desire as these bear on the formation of the multiple and ever-contradictory versions of the "self" and its relationship to the larger society. Hence, pedagogy should provide the theoretical tools and resources necessary for understanding how culture works as an educational force; how higher education connects to other sites of learning; and how identity, citizenship, and agency are organized through pedagogical relations and practices. Rather than being viewed as a technical method, pedagogy must be understood as a moral and political practice that always presupposes particular renditions of what constitutes legitimate knowledge, values, citizenship, modes of understanding, and views of the future. Moreover, pedagogy as a critical practice should provide the knowledge, skills, and culture of questioning necessary for students to engage in critical dialogue with the past, question authority and its effects, struggle with ongoing relations of power, and prepare themselves for what it means to be critically active citizens.

Further, if higher education is to be a crucial sphere for creating citizens equipped to exercise their freedoms and competent to question the basic assumptions that govern democratic political life, academics will have to assume their responsibility as citizen-scholars by taking critical positions, relating their work to larger social issues, and offering students knowledge, debate, and dialogue about pressing social problems. They will also need to provide the conditions for students to have hope and to believe that civic life not only matters, but that they can make a difference in shaping it so as to expand its democratic possibilities for all groups.

In spite of claims made by market fundamentalists that economic growth will cure social ills, the market has no way of dealing with poverty, social inequality, or civil rights issues. It has no vocabulary for addressing respect, compassion, decency, and ethics or, for that matter, what it means to recognize antidemocratic forms of power. These are political issues, not merely economic concerns. A political system based on democratic principles of inclusiveness and nonrepression, in contrast, can and does provide citizens with the critical tools necessary for them to participate in investing public life with vibrancy while expanding the foundations of freedom and justice.

Educators now face the daunting challenge of creating new discourses, pedagogical practices, and collective strategies that will offer students and others the hope and tools necessary to revive the culture of politics as an ethical response to the demise of democratic public life. Such a challenge demands that we struggle to keep alive those institutional spaces, forums, and public spheres that support and defend critical education; help students come to terms with their own power as individuals and social agents; provide the pedagogical conditions for students to learn how to take risks; exercise civic courage; and engage in teaching and research that is socially responsible while refusing to surrender our knowledge and skills to the highest bidder.

In the second section of this book, we focus on the ongoing role that racial politics has played in shaping the liberal arts curriculum, as well as more general questions of how race structures access to and opportunity within higher education. In the first instance, we address the series of intellectual shifts in turn-of-the-century liberal arts curricula from rhetoric to philology to an aesthetic formalism that redefines its object of analysis as universal and race-free. Part of what we hope to demonstrate is that the contemporary call for a "return" to a thoroughly deracinated, formal engagement with the disciplines that make up humanistic study is impossible because race has always been a part of that construction. To further understand the role of interconnected philosophical, political, and social forces on the past and present development of university curricula, we provide in this section a more complicated account of the emergence of literary studies (ground zero of the culture wars) and the simultaneous decline in classical rhetorical study in the academy by situating the institutionalization of this shift in contexts that has been largely ignored or untheorized. We refer specifically to the impact of events such as the rise of social Darwinism and racial science, as well as mass European immigration, which posed significant challenges to the classical liberal principles that have dramatically influenced the role of the university in the production of an active and critical citizenry. In the second instance, we take up the rolling back of educational opportunity and access for minority students at all levels, but particularly in higher education, by focusing on spiraling tuition rates; changes in qualifications for grants and financial aid; challenges to affirmative action; debates over "standards" and testing; and overt attacks on (and defunding of) "politically correct" curricula, programs, and departments—particularly race/ethnic studies. These events are demonstrative of so many strategies of a post–civil rights backlash that threatens not only the civic mission of public and higher education to prepare all citizens for critical participation in self-government, but also undermines any pretense to "freedom" and "equality"—tenets once central to liberal democratic politics.

In the third section, we take seriously both the threat that neoliberalism and corporate values pose to higher education and the necessity to once again remind ourselves that democratic rather than commercial values should be the primary concerns of the university. While the university should equip people to enter the workplace, it should also educate them to contest workplace inequalities, imagine democratically organized forms of work, and identify and challenge those injustices that contradict and undercut the most fundamental principles of freedom, equality, and respect for all people who constitute the global public sphere. Higher education is about more than job preparation and consciousness-raising; it is also about imagining different futures and politics as a form of intervention into public life. In contrast to the cynicism and political withdrawal that media culture fosters, education demands that citizens be able to negotiate the interface of private considerations and public issues; be able to recognize those undemocratic forces that deny social, economic, and political justice; and be willing to give some thought to the nature and meaning of their experiences in struggling for a better world.

If right-wing reforms in higher education continue unchallenged, the consequences will result in a highly undemocratic, bifurcated civic body. In other words, we will have a society in which a highly trained, largely white elite will be allowed to command the techno-information revolution while a low-skilled majority of poor and minority workers will be relegated to filling the McJobs proliferating in the service sector. In contrast to this vision, we strongly argue in the final section of this book that education cannot be confused with training, and that if educators and others are to prevent this distinction from becoming blurred, it is crucial to both challenge the ongoing corporatization of the university and uphold the legacy of a social contract in which all youth, guaranteed the necessary protections and opportunities, once again symbolize the hope for a more equitable and just future. This points to both a democratic project and the need to recapture our commitment to future generations; we must take seriously the Protestant theologian Dietrich Bonhoeffer's belief that the ultimate test of morality for any society resides in the condition of its children. If higher education is to honor this democratic social contract, it not only will have to reestablish its obligation to young people, but reclaim its role as a democratic public sphere.

It is worth noting that the title of this book, *Take Back Higher Education*, should not be confused with the idea of taking over the university, a more militant and authoritarian political concept that we want to avoid altogether. "Take back" is not a call for any one ideology—even a progressive one—to "take over" the university. Our aim is to open up the university to a wider spectrum of debate. But at the same time, we are not shying away from taking a particular stand. "Take back" is an ethical call to action for educators, parents, students, and others to reclaim higher education as a democratic public sphere, a place where teaching is not confused with either training or propaganda, a safe space where reason, understanding, dialogue, and critical engagement are available to all faculty and students. Higher education, in this reading, becomes a site of ongoing struggle to preserve and extend the conditions in which autonomy of judgment and freedom of action are informed by the democratic imperatives of equality, liberty, and justice. Higher education has always, though within damaged traditions and burdened forms, served as both a symbolic and concrete attempt to liberate humanity from the blind obedience to authority and as a reminder that individual and social agency gain meaning primarily through the freedoms guaranteed by the public sphere, where the autonomy of individuals only becomes meaningful under those conditions that guarantee the workings of a democratic society. "Take back" is a reminder that the educational conditions that make democratic identities, values, and politics possible and effective have to be fought for more urgently at a time when democratic public spheres, public goods, and public spaces are under attack by market and ideological fundamentalists, who either believe that markets can best resolve all human affairs or that dissent is comparable to aiding terrorists. Such fundamentalists share a common denominator of beliefs that disable a substantive notion of ethics, politics, and democracy.

Part I

Pedagogy and the Promise of
Democracy in the University

Chapter 1

The Post-9/11 University and the Project of Democracy

A strong military arm of the state, a democratic one in particular, is, as one hears everywhere, the prerequisite for a flourishing economy and a guarantor of its internal order and its sovereignty outside. Even if this were true, it is easy to deceive oneself: military is to democracy as fire is to water. . . . If democracy demands the individual's will, the military demands his subordination. If, in the former case, all power originates from the people, then, in the latter, all orders come from above.

—Ulrich Beck, 1997[1]

We need honest, reasoned debate; not fearmongering. To those who pit Americans against immigrants, and citizens against non-citizens; to those who scare peace-loving people with phantoms of lost liberty; my message is this: Your tactics only aid terrorists—for they erode our national unity and diminish our resolve. They encourage people of good will to remain silent in the face of evil.

—Attorney General John Ashcroft, 2001[2]

September 11, 2001, may well prove a decisive moment in the history of the American university. If, prior to that date, the university was largely understood as a corporate entity whose principle obligation to society was to train a flexible, skilled workforce, in the post-9/11 climate, there seems to be a growing interest in the rhetoric—if not the practice—of civic education, or what it means to teach students to participate as citizens in the moral and political life of a democracy. This renewed interest in its civic mission is largely the result of claims that universities have proven to be the "weak link" in the current war against terrorism. According to this logic, the liberal arts' preoccupation with postmodernism, multiculturalism, and the tenets of tolerance and "cultural relativism" have resulted in a tragic under-emphasis on (or even overt challenge to) the liberal democratic values it is supposed to instill.[3] Although I will claim that the assault on the academy is grievously unfounded, the debates nonetheless open up an opportunity to discuss a number of complex and contested issues, particularly in our moment of crisis, that include: How is democracy variously defined and with what effect? What do calls for civic responsibility and participation specifically demand of citizens? What form of education do citizens require to fulfill such obligations? To what degree do we need to rethink the category of "citizen," given the globalized context in which we now live? In other words, are we merely citizens of a nation-state, or do we require a more cosmopolitan definition of citizenship? To be sure, I do agree with my more conservative interlocutors that it is now "time to teach democracy." Where I depart from their position is around questions of the relationship among democracy, difference, and power—specifically, the university's role in fostering both democratic participation and economic justice. Whereas, for example, conservatives like Diane Ravitch and Lynne Cheney argue that "multiculturalism" has gone too far, I counter that it hasn't gone nearly far enough.

My purposes in this chapter are essentially threefold. First, I want to challenge the aforementioned attacks against the university in the wake of the terrorist atrocities of September 2001. In the popular press, opprobrium has been near universal; charges ranging from moral equivocation to overt anti-Americanism, even pro-terrorist sympathizing have been hurled at college campuses from across the ideological spectrum. In efforts to silence the alleged infamy of those faculty and students who have raised questions about, for example, the integrity of U.S. foreign policy and, more recently, the wisdom of the military occupation of Iraq, the forces united against the "unpatriotic academy" have argued (in a rhetorical sleight of hand that would have astounded Orwell) that now is the time for universities to teach the achievements of Western democracy and not "blame America first" for evils perpetuated worldwide. But what critics as diverse as Lynne Cheney and Todd Gitlin have in mind, I argue, is a pedagogical imperative that more resembles a sacrosanct tribute to a fully realized American democracy than one that creates the conditions for the ongoing political activity of questioning, confrontation, dialogue and dissent central to *democratization*, by which I mean democracy as an ongoing project, *always and necessarily unrealizable*.[4] Second, given this state of affairs, I want to argue that the prevailing campus McCarthyism, which understands itself to be in the service of what we might call for lack of less paradoxical vocabulary, "a militarily secure democracy," is a symptom of the crisis of political democracy itself. This "crisis" includes the redoubled efforts to depoliticize—to militarize and privatize—both public space and public time, for collective deliberation and debate. It also signals the proliferating attacks on difference that have been going on for decades but have accelerated in the current unstable climate. Finally, to deepen and extend a conversation about what the university's responsibility for the institution and perpetuation of political democracy might entail, I offer, in an incomplete and tentative way given the enormity of the task at hand, a series of principles that seem to me an essential part of university education—should it survive—in the new globalized, downsized, neoliberal, post-9/11 world order.

Before turning to the multiple theoretical concerns underlying any claim to "education for democracy," I want to address this question "from the ground," so to speak, by narrating my own experience of teaching in the days following the terrorist attacks. There is little doubt that my pedagogical intervention in the post-9/11 classroom would be perceived as problematic (even heretic) by the Diane Ravitches of the world, so I offer these reflections as "Exhibit A" and leave it to judicious readers to determine what would be lost if we were to abandon our critical capacities or our civic duty to engage the most pressing concerns of our time in the most rigorous, scholarly way we know.

Teaching in the Post-9/11 University

On an unseasonably warm Monday morning in mid-November, almost two months to the day after the September 11 attacks on the World Trade Center and the Pentagon, I walked to my 9 A.M. class filled with apprehension. I was teaching a senior-level American studies course entitled "Hollywood and the Culture of Violence." The principal goal of the course was to expand students' critical vocabulary and complicate their social understanding of the category of violence itself (so that, for example, we might address representational violence or the violence of an ever-consolidating corporate media against a democratic social order, in addition to more routine examinations of violence and individual behavior), the workings of "the culture industry," and the complex relationship between the mass media and society as a whole. We examined a variety of theories of the relationship between Hollywood (and the mass media generally) and violence from fields as diverse as literature, film, sociology, education, women's studies, and philosophy, spanning a period of time roughly from the 1930s to the present. That particular morning, the assigned reading was Douglas Kellner's much celebrated and quite controversial study, "Reading the Gulf War: Production, Text, Reception," an excerpt from his 1992 book-length investigation, *The Persian Gulf TV War*. In spite of the overwhelming and deeply tragic differences between that military engagement and the present occupation of Iraq, there were elements of the strangely familiar that transcended the obvious and gave me pause. Similarities included the controversy over the Pentagon's tight control over journalistic access to all aspects of the military campaign; the White House's official hard line, which paints each war as a conflict (even a "crusade") between freedom- and justice-loving people and "evil-doers" who hate what "we" stand for; the climate of rampant anti-Arabism that it perpetuates; and efforts to represent the American public as united against terrorism (a large percentage of Americans now believing that Saddam Hussein and Al Qaeda were in cahoots, despite all lack of evidence) and in full support of military intervention. (Given the climate of questioning that hangs over the second Bush administration's rationale for a preemptive war against

Iraq and its real motivations, some critical reflection on U.S./Iraqi relations over the past decade now seems even more in order.)

A renowned scholar of postmodern philosophy and culture, media studies, and cultural studies, Douglas Kellner examines the Persian Gulf War in the early 1990s as both a "cultural-political event" and a military one. One of the most successful public relations campaigns in the history of modern politics, he argues, the war in Iraq "can be read as a text produced by the Bush Administration, the Pentagon and the media which utilized images and discourse of the crisis and then the war to mobilize consent and support for the U.S. military intervention."[5] Kellner reads the Gulf War through a methodology he calls "multiperspectival cultural studies," which entails examination of media representations of the war through (1) the production and political economy of texts; (2) textual analysis and interpretation; and (3) audience reception and use of media culture.

What follows from this multiperspectival approach is a detailed investigation of the process of the *production* of news and information that includes analysis of news sources—in this case, primarily the Pentagon and the White House. According to Kellner, high Bush administration officials contacted journalists who would serve as conduits for stories about an alleged Iraqi buildup of over 200,000 troops along the Kuwaiti border threatening to invade the oil-rich kingdom (though satellite photographs purchased by the *St. Petersburg Times* from ABC, who refused to air them, revealed nothing "to indicate an Iraqi force in Kuwait of even 20 percent the size the administration claimed") and the Iraqi unwillingness to negotiate a peaceful settlement in the region (though, again, later transcripts of the infamous Wilson–Hussein meeting failed to support this accusation). Editorial writers and commentators, in turn, used such claims to argue in favor of U.S. military invention as the necessary—and only—effective means of response, in efforts to promote the broader public's total identification with the military point of view. Similar to the current administration's use of "embedded journalists," the first Bush administration was able to further censor and maintain control of the press via the use of a military pool system, which restricted journalists' access to soldiers and the battlefield and thus ensured positive pictures and reporting of the war. Additionally, Kellner exposes disinformation and propaganda, such as the invention of Iraqi atrocities in Kuwait by the U.S. public relations firm Hill & Knowlton, which was later proven patently false on ABC's *20/20* in January 1992. (Similarly, the second Bush administration's stories of aluminum tubes and the like have also been discredited.)

Next, Kellner examines the *text* of the media war: its framing "as a nightly miniseries with dramatic conflict, action and adventure, danger to allied troops and civilians, evil perpetuated by villainous Iraqis, and heroics performed by American military planners, technology and troops."[6] The media repeatedly attempted to personalize the crisis by equating Iraq with its leader, Saddam Hussein, who became the sole actor and source of all evil (over 1,170 news articles linked Hussein with Hitler). The mass hysteria of the American public was whipped up by constant news commentary on the threat of chemical and biological weapons and sudden terrorist attacks, all of which effectively coded the conflict as a Manichean battle between the forces of good and evil and further mobilized support for the war. Kellner also notes the frequency of what Edward Said has described as an "Orientalist" mentality in news coverage of the war, citing as a specific example Jim Hoagland, associate editor and chief war correspondent for the *Washington Post*. Hoagland's prowar position was based largely on the assumptions that Arabs understand only force, are incapable of defending themselves against a much-hated dictator (later it would become clear that Hussein's popular support had been greatly underestimated), and therefore cannot solve their own problems without the aid of U.S. military intervention. In the post-9/11 context, tragically, Hoagland's remarks seem less aberrant than commonsensical. Heads of state such as former Israeli prime minister Binyamin Netanyahu, Italian prime minister Silvio Berlusconi, Harvard University professor Samuel Huntington, and any number of White House officials have described the current global crisis in terms of a clash between the "civilized world" and the "uncivilized world," effectively depicting the terrorist attacks in New York and Washington as a civilizational or cultural phenomenon rather than a political act.

Finally, Kellner analyzes *audience reception* of the 1992 war in Iraq. In no way passive spectators of the media war, U.S. citizens were active in both pro- and antiwar demonstrations and organizing.

Though the media largely ignored the phenomenon, there was a large antiwar movement in place prior to U.S. military intervention, and the country was deeply divided at the start of the war. Nor did the media cover the intense government harassment, intimidation, and mob violence experienced by Arab-Americans during the "Crisis in the Gulf" and the war. Not only were Arab-American businesses bombed, but one Arab-American man was beaten in Toledo, a Palestinian family was shot at while riding in their car in Kansas City, and various activists, including Edward Said and others, received multiple death threats. Kellner notes that harassment had intensified to such a degree during the war that "Pan American Airlines actually decided not to allow Arab passengers on their planes!"[7] What the media did focus on was popular support for the war, in particular the fetishism of yellow ribbons, American flags, prowar demonstrations, and chants of "U.S.A.! U.S.A.!," bolstered by the Bush administration's official line that one was either prowar and thus a good citizen, or antiwar and thus an unpatriotic sympathizer. Though the author voices some alarm at the "quasi-fascistic hysteria unleashed by the Gulf War and the disturbing massification of the public," Kellner is careful not to dismiss public support as the effect of so much mystification or latent xenophobia.[8] "Like sports events and rock concerts," he explains, "the prowar demonstrations . . . provided the participants with at least a fleeting sense of community, denied them in the privatized temples of consumption, serialized media watching and isolated 'lifestyles.'"[9] Participation in an "aesthetic spectacle" of prowar, patriotic ritual thus enabled citizens—typically alienated, anxiety-ridden, powerless—to feel a sense of belonging and power as members of "the winning team in the Gulf War."[10] Kellner makes sense of the open expression of intense militarism, racism, and nationalism in terms of individuals' growing sense of isolation, vulnerability, and insecurity and resultant need to be part of a united, and therefore powerful, community. Further, the nationalism and racism made visible in the public support of the Iraq war (and currently in the war against terrorism) might also be understood as a reaction, in part, to the breakneck globalization of capital and the resulting mass migrations and dislocation of whole populations—events that have spawned fears over the loss of cultural identity and subsequently calls for the tightening of borders and anti-immigrant legislation. At the local level, the community building after the September 11 attacks can also be read as a response to the forces of deregulation and downsizing, and the decline of the welfare state, which have resulted in the withering away of public goods and services and deepened the public's growing sense of extreme individuation and alienation.

In the comparative halcyon days of pre-9/11 July, when I planned the fall course, making the Kellner selection required reading for a course on media and violence seemed like a very good idea. Kellner's study of the use and abuse of media in the Gulf War raised fundamental questions about the extent to which corporate and governmental power might encroach on a "free" media, the degree to which we rely on its capacities to be rigorous and critical so that we might effectively fulfill *our* duties as informed citizens, and the high-stakes struggles for such freedoms in times of war, given the implicit tensions between democracy and militarism.

Walking to class that morning, admittedly, I doubted not so much my convictions, but rather my timing. I had witnessed my students' responses to the terrorist attacks of September 11, sat with them as we watched CNN together—ripped from the reassuring commonplaces of our everyday lives, of classes and readings and discussions, unable to focus on anything else. "Why us? Why do they hate us?" they demanded. In the days immediately following the catastrophe there were many conflicting answers to these pressing questions. Was it a result of U.S. support for antidemocratic regimes in the Middle East? Its military presence in the Holy Lands? Its near unilateral support of Israel over Palestine? A hatred of American freedoms? A clash of civilizations? A holy war? Then, two months after the attacks, I was responding to their confusion and fear—largely undiminished due to nightly terrorist alerts punctuated by ongoing discussions of biological/chemical/ nuclear warfare—not with assurances of a happy ending, the return to life as we knew it, or the promise of a swift and just resolution for all. (How could I? How could anybody?) To counter their sense of helplessness, I offered what I could—an opportunity for them to use their critical capacities to participate in a public political sphere, to doubt, to raise questions, to assent or dissent as their consciences guided them, with the implicit understanding that "no problem is resolved in advance. We have to create the good, under imperfectly known and uncertain conditions."[11] But while the use of one's reason remains a precondition for agency, it flouts, as Zygmunt Bauman argues, "the desperate human desire for reassurance."[12] On one hand, the official rhetoric of the White House is better equipped to quench a frightened public's thirst for certainty ("we" are friends, "they" are our enemies), absolute foundations ("we," the good, love freedom and democracy, "they" are evil and hate both), and codes of practice (be patriotic, spend money in New York, defer to military experts, do not ask questions about things you couldn't possibly understand)—it is far better than the "freedom-cum-uncertainty cocktail served by autonomous reason."[13] On the other hand, doubts can also liberate. Contrary to widespread opinion, doubt may not be the last refuge of postmodernists and other unpatriotic scoundrels, but a condition that, as Ulrick Beck argues, "makes everything possible again: questions and dialogue of course, as well as faith, science, knowledge, criticism, morality, society, only differently, a few sizes smaller, more preliminary, revisable and more able to learn."[14]

And there is reason, everywhere, to doubt. The sense of the uncanny that Douglas Kellner's analysis invokes is difficult to ignore—the Gulf TV war itself a bizarre Mobius strip translating simulation into reality and back again—and is now repeated and doubled in hideous nightmare proportions, a product of Hollywood's demonic imaginary returned as reality. Only this time we didn't begin with a simulated buildup of troops along the Kuwaiti border, but with the unreal reality of four passenger airplanes transformed into instruments of apocalyptic destruction, the ruin of lower Manhattan, the savaging of the Pentagon, the blackened crater in the bucolic farmlands of southern Pennsylvania, the scenes of a makeshift burial ground for some 3,000 innocent civilians. This was no "media event" to be sure. But in the days after the crisis, we would all rely on media to keep us informed—who has perished? Who has survived? What might we do to help the families of victims? Was it safe to fly? Were we safe in our homes? How were we coping? How was life irrevocably changed? Would we go to war? Would that stop terrorism? Why, why, why did this happen? Needless to say, there is no justification for such horrific devastation. But at the same time, acknowledging that we have been victims doesn't alleviate the burden placed on us to come to terms with what has happened and the necessity of response, both individually and collectively. Given all that is at stake in an ever-advancing war on terrorism, we cannot divest ourselves of the responsibility to critically examine the history of U.S.-Arab relations as well as those conditions (historical and political as opposed to simply and irreducibly cultural) that produce terrorists. This requires, in addition, interrogating how a global media—controlled by a handful of multinationals—frames such histories for us, as well as the interests and purposes they serve. If, as media theorist Nick Couldry has recently argued, the events of September 11 can be read as "a violent challenge to a world where symbolic inequality parallels and reinforces other kinds of inequality"—in particular the extreme unevenness of the world's media operations—then the prevention of future atrocities demands immediate attention to the possibilities of democratizing the media landscape to create real global dialogue.[15] Yet, I wondered: would my students perceive the assignment as unpatriotic? Inflammatory? Insensitive? I had a young woman in class who has two brothers—one employed in the World Trade Center and one at the Pentagon—both at work the morning of 9/11 and both miraculously survivors. Was it possible that the analysis of the Gulf War as a deeply problematic moment in the history of U.S.-Arab relations could be conceived as offering "justification" for the civilian attack? Would they think I was implicitly suggesting to the family member of two victims of the terrorism that they/we somehow "deserved it" or "had it coming"?

My students gave no indication they read it that way. In contrast to my own dishevelment upon arrival that morning, they were composed, serious, and ready to work. Any worry I had that this was an inopportune or inappropriate time for discussion of events related to the recent trauma they had experienced quickly dissipated. Students seized the opportunity to work through the ocean of conflicting commentaries on the terrorist attacks and the war in Afghanistan that had washed over them since 9/11. Though the media representation of the Gulf War and America's current wartime operations offered many points of comparison, the distinct differences between that historical context and the current one were hardly lost on them. We were the victims of an outrageous attack, and they all agreed that no history of bad blood justified the loss of so many thousands of innocent lives. Implicitly they understood the distinction between explanation and justification, apparently lost on former press secretary Ari Fleischer, Defense Secretary Donald Rumsfeld, and Senator John McCain and members of the Bush administration who warned Americans that they "must watch what they say."[16] And explanation they desperately needed. Having outgrown fairy tales, fantasy, and epic sci-fi of the *Star Wars* variety, they were dissatisfied with an official rhetoric of good versus evil that neither aided their efforts to achieve a more complex accounting of the global crisis in order to engage it in the interests of a more just and secure future, nor provided them with a language for uniting victims of violence around the world who are forced to live under conditions that undermine their rights, their dignity, and their humanity.

Students also discussed the way the world had changed in a decade since the infamous "television war." From their perspective, Kellner's analysis of the tight control that the elder Bush administration had on representations of the Gulf War was no longer possible in a wired world, in spite of the Pentagon's efforts to manage the public's reception and understanding of the war in Afghanistan (and the more recent war in Iraq). All savvy users of the Internet, they had access to a mass of opinion whose ideological range was matched by its geopolitical reach. They knew how to access alternative news sites in the United States as easily as they could read the *Guardian*, or for that matter leading Arab dailies. Indeed, many sites, they were eager to point out, archive international news articles on all events relating to 9/11. Further, perhaps as a result of the controversy that surrounded the Gulf War, mainstream newspapers like the *New York Times* self-consciously examined the deepening crisis in Afghanistan from the perspective of a "highly orchestrated communications" event.[17] What ensued from that point on were debates about modern warfare as a reasonable and effective response (given the problematic conflation of terror cells with host countries) and steps taken by the current administration to curb civil liberties in the name of national security—this was before John Ashcroft's inflammatory and reckless equation of those who critique the president's administration with the terrorists themselves. The upshot of their investigations into the Gulf War and by implication the current war on terrorism was not a desire to "blame America" or denigrate democratic ideas, but neither was there a desire to resurrect a sacrosanct image of America the Blameless. Rather, by recognizing all that was at stake, the exercise encouraged them to take a more active and critical role in their own reception of war information, representations of U.S. foreign relations, national political rhetoric, and the way their understanding of these issues informed their participation in public debates. Further, our discussion provided them with a vocabulary for engaging "the enormous inequality in which voices, and even which regions, contribute to the truly global flows of images and narratives"—an effective point of entry into decades of debate about the global media industry's complicity with forms of cultural imperialism.[18]

So there I was at the end of class doubly vexed. First, having spent the better part of a semester teaching my students to be canny about the ways media manipulate public opinion and manufacture consent, I was the one—not my students—who bought the "official line" about patriotism, at least for a little while. Second, I had really sold my students short. There was no need to defend a critical reading of U.S.-Arab relations against charges of "anti-Americanism"; they understood that political debate is central to the core commitments of a democratic citizenry to shape their own laws and public policy and, when necessary, to change them.

The Weakest Link in the War on Terrorism?

By the time I returned to my office from class, a friend had e-mailed me a story from the *Chronicle of Higher Education*. According to the article, the American Council of Trustees and Alumni (ACTA), a conservative academic watchdog group founded by Lynne Cheney and Joseph Leiberman, issued a report imperiously titled, "Defending Civilization: How Our Universities are Failing America and What Can Be Done About It." The authors, Jerry L. Martin and Anne D. Neal, proclaimed that while citizens across the country—92 percent according to one poll—had rallied behind the president and favored the use of military force against Afghanistan after the September 11 attacks, "college and university faculty are the weak link in America's response to the attack"—and they named names.[19] In fact, there were 117 names of scholars, students, and even a university president singled out for unpatriotic behaviors gathered from student newspapers, web sites, and the media and compiled in a manner reminiscent of McCarthy-era blacklisting. Wagging their fingers at an unpatriotic academy overrun by equivocating moral relativists, tolerance-mongering multiculturalists, and left-wing terrorist sympathizers, the authors excoriated faculty members who "refused to make judgements," "invoked tolerance and diversity as antidotes to evil," and "pointed accusatory fingers, not at the terrorists, but at America itself."[20] Such acrimony aside, most of the statements they cited were innocuous. Consider, for example, Jerry Irish, professor of religious studies at Pomona College, who states, "We have to learn to use courage for peace instead of war." Or Stanford University professor Joel Beinin, who argues "If Osama Bin Laden is confirmed to be behind the attacks, the United States should bring him before an international tribunal on charges of crimes against humanity." Or Todd Gitlin, professor of communications at New York University, who suggests that "There is a lot of skepticism about the administration's policy of going to war." Juxtaposing commentaries provided by scholars who voiced concern over the administration's decision to go to war or urged a rethinking of U.S. foreign policy with those of President Bush ("In this conflict, there is no neutral ground"), Mayor Giuliani ("You're either with civilization or the terrorists"), and leaders of both the Democratic and Republican parties, the authors of the report formulated the following conclusion. "Rarely did professors publicly mention heroism," they asserted, "rarely did they discuss the difference between good and evil, the nature of Western political order or the virtue of a free society. Their public messages were short on patriotism and long on self-flagellation. Indeed, the message of much of academe was clear: BLAME AMERICA FIRST."[21]

Not only has the ACTA charged the academy with rampant anti-Americanism, but also with creating a climate hostile to students and professors, particularly if they are untenured, who support the

war efforts first in Afghanistan and then Iraq, but feel too
browbeaten by peaceniks to display their patriotism publicly. Only
those who oppose the war on terrorism are licensed to voice their
opinions and concerns. The fear of reprisal felt by Bush's supporters has, from ACTA's perspective, a chilling effect on "the robust
exchange of ideas . . . essential to a free society."[22] But their arguments remain dicey at best. Attempting to appropriate heretofore
progressive slogans in this war of position, they suggest that conservative students and faculty are being actively denied their right
to dissent, their right to assemble, and their right to free speech.
Though government, military, and popular support clearly stand in
their favor—and no one has posted their names, campus location,
and accompanying commentary on the Web—they are nonetheless
victims of ruthless intimidation. Further, the ACTA claims to support the guarantees of freedom of speech for all, yet they insist
"*Academic freedom does not mean freedom from criticism*," a
warning they direct at those who would question U.S. foreign policy in a time of war. Who is intimidating who? While we must all
acknowledge that the positions we hold bear consequences, it
seems that the ACTA wants to make sure that those who dissent (I
use the word here in its more traditional sense of speaking truth to
power, as opposed to speaking power's version of truth) feel the
real weight of their choices. Of course, there is more at stake here
for members of the ACTA than making critics of American foreign
policy accountable for their positions. As Curtis White argues,
"For this group, any movement away from the grossly patriotic is
a 'failing.' They would save American culture by removing from it
any thought that isn't utterly conformist with the opinion of 'the
public at large.'"[23]

The ACTA's answer to this apparent campus pandemic of politically correct anti-Americanism is to urge the 3,000 trustees at colleges across the country to whom the report was sent to apply
pressure to "adopt strong core curricula that include rigorous,
broad-based courses on the great works of Western civilization as
well as courses on American history, America's Founding documents,
and America's continuing struggle to extend and defend the principles
on which it was founded."[24] Further, they urge that "If institutions
should fail to do so, alumni should protest, donors should fund

new programs, and trustees should demand action."[25] To underscore the significance and urgency of this appeal, they cited support from the White House. Founder of the ACTA and wife of Vice President Dick Cheney, Lynne Cheney argues,

> At a time of national crisis, I think it is particularly apparent that we need to encourage the study of our past. Our children and grandchildren—indeed all of us—need to know the idea and ideals on which our nation has been built. We need to understand how fortunate we are to live in freedom. We need to understand that living in liberty is such a precious thing that generations of men and women have been willing to sacrifice everything for it. We need to know, in a war, exactly what is at stake.[26]

What such a historical engagement might look like is the subject of her 1995 book, *Telling the Truth: Why Our Culture and Our Country Stopped Making Sense and What We Can Do About It*. Preeminently an attack on multicultural education in general and National History Standards in particular, Cheney asserts that such curricula encourage "students to take a benign view—or totally overlook—the failings of other cultures while being hypercritical of the one in which they live."[27] To cement her case, she argues that as a result of current multiculturalist reforms, students are taught a great deal about the Ku Klux Klan and Sen. Joe McCarthy, but little about the scientific and technological achievements of such figures as Alexander Graham Bell, the Wright brothers, Thomas Edison, Albert Einstein, and Neil Armstrong. The results, for Cheney, are disastrous; she cites a number of multicultural educators whose intentions she describes as building a case against patriotism. In times of peace, multiculturalism produces nothing short of the ideological disuniting of the nation, and in times of global conflict, it produces an increased unwillingness to wage all-out war. "As American students learn more about the faults of this country and the virtues of other nations," she argues, "they will be less and less likely to think the country deserves their special support. They will not respond to calls to use American force. . . . "[28] Clearly, the exercise of critical reflexivity before putting the nation's young men and women in harm's way is not a virtue, according to this logic.

I wondered what my students would have done if I'd intervened in our discussion of war and the role of the media in shaping public opinion to remind them of great American contributions to science and technology, of heroes like Alexander Graham Bell and the Wright brothers.

In an essay that appeared a month after the terrorist attacks, entitled "Now is the Time to Teach Democracy," Diane Ravitch echoes Cheney's sentiments and takes the argument a few steps further. She argues that multicultural education not only promotes disunity, but is dangerous to democracy because it encourages people to think about "our own racial and ethnic differences," and advances not tolerance but a rejection of the common good and the erosion of civil society. Further, she challenges (in a move that suggests a growing irrational, if not hysterical, mood) calls to teach tolerance in the aftermath of September 11, "as if our children were somehow responsible for what happened because their teachers had failed to teach them tolerance."[29] It soon becomes clear that what is underlying this dubious logic is the fear of implication, hence her assumption that according the attacks some rationality by situating them in a context of U.S.-Arab relations and locating the conditions that produce extremism would necessarily lead to a sympathetic identification with terrorists and a justification for their actions. She writes:

> Now, in the wake of the terrorist attacks, we hear expressions of cultural relativism when avant-garde thinkers tell us that we must try to understand why the terrorists chose to kill thousands of innocent people, and that we must try to understand why others in the world hate America. Perhaps if we understood why they hate us, then we could accept the blame for their actions.[30]

Surely it is possible to condemn terrorism and innocent human suffering and to probe, at the same time, the conditions that gave rise to it. Would it be reasonable, let alone responsible, to call for the eradication of AIDS and then refuse to address those conditions—behavioral as well as economic, cultural, and political—that allow the virus to thrive on grounds that we are "blaming the victims"? Efforts to end terrorism or cure AIDS demand critical reflection and relentless questioning—how would it be possible otherwise to build a future without fear or imminent threat of brute violence or widespread disease? Judith Butler argues that

> Our collective responsibility not merely as a nation, but as part of an international community based on a commitment to equality and non-violent cooperation, requires that we ask how these conditions came about, and to endeavor to recreate social and political conditions on more sustaining grounds. . . . Can we hear at once that there were precedents for these events, and to know that it is urgent that we know them, learn from them, alter them and that the events are not justified by virtue of this history and that the events are not understandable without this history?[31]

It seems, however, that the only social and political conditions for the terrorist atrocities of 9/11 that conservatives demonstrate any patience for unearthing are those they shakily connect to "left" intellectual discourses circulating within the university. In the face of this massive threat to Western law and order, it seems, questions of U.S. foreign policy, with its nearly unconditional support of Israel, its alliances with undemocratic even despotic governments in exchange for petro-dollars, and its military occupation of Holy Lands, are relegated to the proverbial dustbin of history, as are questions of U.S. control of the global media landscape, the symbolic inequalities that issue from this, and the vast material inequalities suffered by the region. Consider, for example, an editorial in the *New York Times* published on September 22, not two weeks after the attack, by Edward Rothstein. Dismissing what he calls "ideological confections" such as "corporate profit-taking and political power" or worldwide protests against "globalization," Rothstein quickly uncovers the real culprits in the destruction of the World Trade Center and the Pentagon—postmodernism and postcolonialism. In his analysis, postmodernism is simply a "challenge" to "the assertion that truth and ethical judgment have any objective validity," which apparently denies intellectuals any ethical criteria for evaluating the practices of other cultures, and postcolonialism "focuses on cultures that have experienced Western imperialism" in ways that lay the groundwork for theorizing—and it seems forgiving—all acts of terrorism as simply responses to said imperial behavior.[32] Throwing all scholarly caution and integrity to the wind with such wild reductions, Rothstein concludes:

> One can only hope that finally, as the ramifications sinks in [sic], as it becomes clear how close the attack came to undermining the political, military and financial authority of the United States, the Western relativism of pomo and the obsessive focus on poco will be widely seen as ethically perverse. Rigidly applied, they require a form of guilty passivity in the face of ruthless and unyielding opposition.[33]

Similarly, the *Weekly Standard* ran an editorial screed by Waller R. Newell in late November 2001 entitled: "Postmodern Jihad: What Osama bin Laden learned from the Left." Like Rothstein, Newell traced—in a bizarre juxtaposition of intellectual traditions—the reigning political ideology of Al Qaeda to "European Marxist postmodernism," and in particular to the work of Martin Heidegger, Jean-Paul Sartre, Frantz Fanon, and Jacques Derrida. Like his U.S. interlocutors, Newell has ratcheted up the rhetorical wattage of his critique of the "liberal" university, assuming, it seems, that such animus provides adequate compensation for his apparent lack of scholarly interest in the actual content of much late-twentieth-century philosophical thought. He roughly charges, "[Postmodernism] has damaged liberal education in America. Still it doesn't kill people—unlike the deadly postmodernism out there in the world. Heirs to Heidegger and his leftist devotees, the terrorists don't limit themselves to deconstructing texts. They want to deconstruct the West, through acts like those we witnessed on September 11."[34] Feeling, perhaps, that he had let "postmodern" academics off the hook with this last concession, he quickly recovers himself, assuring that "What the terrorists have in common with our armchair nihilists is a belief in the primacy of the radical will, unrestrained by traditional moral teachings such as the requirements of prudence, fairness, and reason."[35] Prudence, fairness, and reason, however, read like a hollow banner for conservatives armed with absolutes, consistently brandishing the "you're with us or you're complicit with terrorism" charge like a call to arms in yet another fundamentalist crusade.

Finally, the case of the young "American Taliban" John Walker provided conservatives with clear proof that a formal and informal education steeped in "multiculturalism" and "postmodern relativism" not only incited pathology and violence in the barely civil East, but had profoundly deleterious effects on "our own" naïve youth. When CNN's *People in the News* profiled Walker, it was clear the kid never had a chance. A product of a "high income, ultra liberal" boho Northern San Francisco community and child of divorced parents, the teenage Walker "immersed himself in rap music, listening to artists like L.L. Cool J. He would often visit hip-hop Web sites, once even posing in an email as an African-American. Identifying with other ethnic groups and cultures," *People in the News* host Daryn Kagan ominously intones, "would become an important characteristic of John Walker's."[36] Later, we are told that "Walker's father says the turning point may have come when his son read *The Autobiography of Malcolm X*."[37] The scene cuts to Shelby Steele of the Hoover Institute, who theorizes Walker's investment in African American cultural production in the following terms: "The way to be hip, the way to be cool, is to take on a little theme of anti-Americanism, to identify with things from other cultures, to identify with black alienation."[38] Not only does Steele deny the validity of black alienation by reducing it to a form of anti-Americanism, but he also critiques those who are in sympathy with the protest traditions of black Americans—which have historically addressed fundamentally democratic concerns with social justice, racial equality, freedom, and self-determination—as a shortcut to "cool." The political and pedagogical implications of CNN's carefully crafted packaging of the Walker story are quite clear: learning about and identifying with other (nonwhite) cultural and political traditions is dangerous, while education in "universal" (mainstream, white, American) values and traditions is both patriotic and pure. Conservative commentator Ann Coulter capitalized on the John Walker story by using it to suggest that Walker was the poster boy for the sins of liberalism and that executing him could serve a useful political purpose—one that aligns her more closely with Osama bin Laden than anyone else. She writes: "When contemplating college liberals, you really regret once again that John Walker is not getting the death penalty. We need to execute people like John Walker in order to physically intimidate liberals, by making them realize that they can be killed too. Otherwise they will turn into outright traitors."[39]

Yet the unwillingness to take up the challenge of what Butler calls a "new order of responsibility" is pervasive not only among conservatives, but among many progressives as well. Consider, for example, the comments of Todd Gitlin in an issue of *Mother Jones*. At first, Gitlin calls for a "reasoned, vigorous examination of U.S. policies, including collusion in the Israeli occupation, sanctions against Iraq, and support of corrupt regimes in Saudi Arabia and Egypt," as well as "the administration's actions in Afghanistan and American unilateralism."[40] However, in the public debates that followed the tragedy of September 11, Gitlin argues, what we have encountered is not responsible criticism, but so much "left-wing fundamentalism, a negative faith in America the ugly."[41] In a specious distinction reminiscent of John Ashcroft's early December 2001 speech before the Senate Judiciary Committee equating dissent with terrorism, Gitlin juxtaposes a kind of "soft anti-Americanism that, whatever takes place in the world, wheels automatically to blame America first" against the "hard anti-Americanism of bin Laden, the terrorist logic under which, because the United States maintains military bases in the land of the prophet, innocents must be slaughtered and their own temples crushed."[42] Who is being reductive in this instance? While there is no justification for bin Laden's acts of terrorism against the United States, the grounds for his actions are not reducible to the U.S. military presence in Saudi Arabia. And Gitlin himself knows better, given his own insistence that we must rethink U.S.-Arab relations and the current military campaign. Moreover, he offers support for his charges of "soft anti-Americanism" in the form of brief quotations, pulled from their contexts, from Edward Said, Noam Chomsky, and Arundhati Roy—each an acclaimed scholar or writer whose contributions to the study of U.S. foreign policy in the Middle East can best be measured in years if not decades. That his proofs are feeble or unfair, however, is not really the issue. In a rhetorical sleight of hand, Gitlin has successfully lumped together the voices of dissent with the actions of terrorists—one big melange of anti-Americanism—and made a mockery of the kind of democratic debate to which he initially appeared to subscribe.

The 1991 Gulf War is not the only occasion for collective déjà vu—the current charges of academic anti-Americanism offer yet another blast from the past. The acrimony and venom heaped on the university since the terrorist attacks suggest that while much has changed in American life since the catastrophic events of September 11, the debates over liberal arts education—ongoing since the early 1980s—remain unrefreshingly unchanged. To be sure, now indeed is "the time to teach democracy" in American universities. But what does it mean when "democracy" can be invoked by conservatives to lend credence to the dictates of a rigid authoritarianism that pits patriotic, white Americanism against critical multiculturalism? Or against radical educators, who support the education of citizens as citizens who are capable of making their own laws and when necessary changing them. What does it mean when teaching "democracy" becomes so wedded to the dictates of market capital that the university now understood as corporate entity can eschew its civic mission, however defined, in pursuit of the logic of pure market? Clearly the crisis of vision and purpose that marks the post-9/11 university cannot be separated from the broader crisis of democracy, to which I now want to turn.

The Crisis of the University Is the Crisis of Democracy

In the 1997 essay written just before his death, Cornelius Castoriadis delivered a devastating critique of the current situation of most so-called democracies of the West, in which neoliberal social and economic policies were consigning political institutions and political agency to insignificance. "Democracy as Procedure and Democracy as Regime" reveals the vast distance between the governing structures of contemporary U.S. society (what he calls a "procedural democracy") and its ideal of political democracy. In a procedural democracy, citizens are effectively removed from political choice, performing the duties of citizenship such as voting—typically on issues about which they have had little say—once every four years, out of a sense of compulsion or routine. The atrophy of democracy in this sense betokens the ascendancy of neoliberal social and economic policies "whereby a relative handful of private interests are permitted to control as much as possible of social life in order to maximize their personal profit."[43] Neoliberalism appears to "work best under conditions of procedural democracy, when the population is diverted from the information, access, and public forums necessary for meaningful participation in decision making."[44] This becomes clear in the current moment of crisis as public discussion about the integrity of U.S. foreign policy in the Middle East, and especially Iraq, is coded as being in sympathy with terrorism by high-ranking officials in the Bush administration. In this instance, the interests of corporate profit and the interests of militarization inform and advance one another.

No longer interpellated as members of a political community at the national or local level, individuals seek affiliation and belonging elsewhere—in family life, in religious or ethnic communities, or through participation in the patriotic aesthetic afforded by popular culture. As agency gets reduced to lifestyle decisions, making wise consumer choices, and agonizing over private troubles within the context of neoliberal economics and its relentless attack on all things social, Zygmunt Bauman argues that the meaning and experience of "public life" and "public interest" undergo a radical transformation. The public forum, a site where the collectivity discusses and debates those concerns and interests that arise out of the perennial quest for democratic self-governance and its ongoing obligation to transform the present in the service of a more just future, has been replaced by the "talk-show-style surrogate" of the forum, to use Bauman's phrase. He explains:

> The public—the gathering of other individuals—can only applaud or whistle, praise or condemn, admire or deride, abet or deter, nudge or nag, incite or dampen; it would never promise to do something that the individual could not do herself or himself, to tackle the problem for the complaining individual (being but an aggregate of individual agents, the listening/commenting public is not an agency in its own right), to take the responsibility off the individual's shoulders.[45]

Without a collective investment in the alleviation of the pain and suffering of its individual members, the individuals who make up society are made to understand that their burden is theirs alone to endure or not. In this sense, the "public interest" becomes a metaphor for that which titillates and amuses, shocks and frightens, while "the public" plays the role of spectator—cheering or harassing on cue without ever entertaining the notion of intervention. As Bauman's comments make clear, neoliberalism's ongoing efforts to privatize and depoliticize the commonweal is not only about privileging the alleged freedoms of every one of its members to pursue their individual interests and satisfactions. It is also about the loss of political agency and effectiveness; it is about dismantling the welfare state and revoking its obligations to make the lot of those private individuals easier to cope with through investment in public education, health care, and adequate housing. In an era of privatization, Margaret Thatcher once infamously asserted, "there is no such thing as society," only an agglomeration of individuals and their private interests. Moreover, in the aftermath of the terrorist attacks of September 11 and the recent occupation of Iraq, citizens racked with fear have assented to the stepped-up militarization of daily life and the rolling back of civil liberties in the name of "national security," and at the same time further diminished their individual and collective capacities to actively participate in democratic life. In an era dominated by neoconservatives and market fundamentalists, the "consensus on defense is the 'antidote' to the democratic consensus," making it possible for governmental powers to "dismiss democracy with the blessing of democracy."[46]

How, then, to revitalize the public sphere under the current aggressive conditions? How to transform a "procedural" democracy into a substantive democratic "regime" of autonomous individuals living in an autonomous society? How to challenge the forces of militarization pitted against the civil liberties that vouchsafe our critical capacities as citizens? Obviously, a democratic regime requires individuals capable of not only bringing it into being, but also making it function and reproducing it. Hence, Castoriadis asserts, "the question of *paideia* proves ineliminable."[47] A prodigious political educational process, or *paideia*, must be committed to developing and exercising all those abilities that democracy requires, so that its claim to autonomy and political equality is "as close to the effective reality of that society as possible."[48] While it is crucial to understand the possibility of a truly democratic *paideia* as imperiled, it is equally important to insist that it is not yet beyond our grasp, as we take measure of a radically diminished public sphere and the vulnerability of public institutions such as higher education.

Castoriadis's distinction between democracy as procedure and as regime demands a response to what I would argue is the central issue for scholars in the academy: How does our work encourage or undermine the civic capacities of those we attempt to educate? Jacques Derrida elaborates on this problem by posing the following questions:

> Who are we in the university . . . ? What do we represent? Whom do we represent? Are we responsible? For what and to whom? If there is a university responsibility, it at least begins with the moment when a need to hear these questions, to take them upon oneself and respond, is imposed. This imperative for responding is the initial form and minimal requirement of responsibility.[49]

How we choose to answer these questions has dramatic implications not only in terms of the impact we will have on the fate of the university, but also for the kind of society in which we live. It is a curious contemporary circumstance that many progressive academics will describe their work as political, even radically transformative, and yet refuse to name, however provisionally, the kind of political regime they hope to put into place, or even acknowledge the ethics that drive their political commitments, alliances, and decisions.

In response to the wartime imperative to "teach democracy now," I want to provide a rationale for the important and necessary use of the democratic imperative to expand individual and collective capacities to self-govern as an ethical referent for what we in the university do. Such a project is necessary not only to defend in the most immediate terms higher education—particularly the humanities—against its ongoing depoliticization and vocationalization, but also, more crucially, to defend it as one of the last public spheres for the open dialogue and debate so essential to the ongoing political education of citizens and the revitalization of democracy itself. Defining the university as a vital public sphere requires a commitment not only to technical, professional training, but also to critical and ongoing political education in the service of democratic social relationships and participation in public life. If such participation necessitates public space, "Only the education (*paideia*) of the citizens as citizens," Castoriadis argues, "can give valuable, substantive content to the 'public space.'"[50] Hence the question of meaningful public space for discussion, debate, and compromise—as opposed to the current space of mass-mediated schizophrenia promoting a violent juxtaposition of consumer fantasy and the culture of fear, from Britney Spears peddling Pepsi to the eerie green night vision of U.S. troops in Iraq, from mass advertising to mass militarization—cannot be separated from the obligations of *paideia*. And equally important is the creation of a "public time," which Castoriadis defines as "the emergence of a dimension where the collectivity can inspect its own past as the result of its own actions, and where an indeterminate future opens up as domain for its activities."[51]

The project of engendering a genuinely democratic society and individuals capable of sustaining it is, of course, a fraught proposition in several respects. The initial dilemma concerns the vulnerability and the potential mortality of a democratic regime, as a regime that functions by virtue of critical reflection and self-government, once citizens effectively subject such institutions to scrutiny, reexamination, critique, and evaluation. Either citizens take part in an alleged form of democracy in which they simply "apply the procedures" in ritual fashion, or they recognize and act upon the need to become self-governing. Accordingly, the civic education of individuals will be dogmatic, authoritarian, and passive, or it will be critical—and then, as Castoriadis argues, "the Pandora's box of putting existing institutions into question is opened up and democracy again becomes society's movement of self-institution— that is to say a new type of regime in the full sense of the term."[52] I highlight the indeterminancy and openendedness associated with the institution of a regime of political democracy to counter the often repeated warnings by many academics who are critical of appeals to a "modernist" language that posits democracy as a referent for ethical action. Though certainly there are reasons to be cautious of its utopian elements, I hold that certainty and closure are not necessarily among them.

The risk of the institution of a new type of regime and the demise of democracy is only one problem. The axiom that an autonomous society, a society that makes its own laws, must be made up of autonomous individuals (and that individuals can only be autonomous in an autonomous society) also demands a radical rethinking of political philosophy in general and democratic theory in particular. It requires rejecting the model of the separation between the public and the private domains and recognizing instead their *mutual dependence*.[53] And, relatedly, it means challenging the myth of individualism, the autonomous self who can exist apart from society, as well as reified notions of the social institutions that comprise "the State," which are often conceived as in opposition to "the people" rather than as being constituted by them. The question is how to rethink the crisis of politics and the possibilities for a genuinely democratic *paideia* in light of these challenges to core assumptions in political theory.

The ancient Greeks made a distinction among three spheres of human activity that the overall institution of society must separate and articulate: the *oikos*, the *agora*, and the *ekklesia*. According to Castoriadis's translation these are the private sphere, the private/ public sphere, and the public sphere, respectively. The *oikos*—the family household, the private sphere—is where (at least formally and in principle, Castoriadis reminds us) political power neither can nor should intervene. *Ekklesia*—the governing body where matters affecting members of the *polis* are addressed—is the site of effective political power. Traditionally these have been understood as *separate* and mutually *independent* spheres, but as Bauman rightly argues, it is "rather the link, the mutual *dependence*, the *communication* between the two domains which should lie at the center" of political and specifically democratic theories.[54] The third, in-between space of translation is the *agora*, which binds the two extremes and holds them together. By focusing on this third sphere of action, Castoriadis develops a theory of social and political organization that at the same time challenges the binary "civil society/State" distinction that has been the hallmark of contemporary political thought.[55] The *agora* is crucial to the maintenance of a genuinely democratic regime, of an autonomous *polis* made up of autonomous members. "Without it," Bauman insists, "neither the *polis* nor its members could gain, let alone retain their freedom to decide the meaning of the common good and what [must be] done to attain it."[56]

However, there are two ways, Bauman continues, in which the *agora* may be jeopardized, "its integrity endangered and its role distorted or altogether undermined with the effect of folding back the autonomy of society and its individual members alike."[57] One is old and familiar; the twentieth century witnessed the rise of several totalitarian regimes worldwide. The totalitarian project seeks the total annihilation of the private sphere, while the public sphere is not at all public in the sense that citizens have access to it, but rather exists as the property of the totalitarian state. The other is quite new and, as I have attempted to show above, currently unfolding in the contemporary societies of the West, where neoliberal economic and social policies have achieved nearly the reverse phenomenon: the collapse of the public sphere into the private. It is as if, living in the shadow of twentieth-century totalitarianism, Westerners have learned their lessons very thoroughly: state control is corrupt, so better to turn everything over to the free market. Address most Americans on the subject and this is likely the response. To the degree that the state is perceived as a threat to personal freedoms rather than a guardian of the public good, the laws of the market supersede the laws of government, and the freedom once vouchsafed as a universal right for all members of society is literally cashed in for the freedom of each individual member—to consume, to make profit, or struggle to survive *alone*. With the advent of Reaganism in the 1980s this antistatist tendency was exacerbated by a carefully crafted Republican rhetoric that linked "big government" with the interests of minorities, women, and gays. The increasing marketization and privatization of everyday life has come at the cost of notions of community and the common good. As Bauman insightfully argues, "The 'public' has been emptied of its own separate contents; it has been left with no agenda of its own—it is now but an agglomeration of private troubles, worries, and problems."[58] In some respects, the attacks of September 11 initially produced a revival of support for the federal government and with that the construction of a new Office of Homeland

Security. But curiously, this office has nothing to do with security in the more mundane "homeland" sense of providing jobs, adequate health care, child care, or a living wage—and everything to do with the security of markets and military might. Moreover, two years after 9/11, big government as a provider of social services (except for those organized around corporate welfare) is once again disparaged and discredited by those in power.

Any attempt to recuperate political agency, then, must begin with a profound reorientation of society, with the recapturing of the *agora* from commercial, financial, and military abuse. "To make the *agora* fit for autonomous individuals and autonomous society," Bauman asserts,

> one needs to arrest, simultaneously, its privatization and depoliti-
> cization. One needs to re-establish the translation of the private into
> the public. One needs to restart (in the *agora*, not just in . . . semi-
> nars) the interrupted discourse of the common good—which renders
> individual autonomy both feasible and worth struggling for.[59]

In this way, the *agora* might become effectively public, "an *ekkle-sia* and not an object of private appropriation by particular groups."[60] The necessary rebuilding of the *agora* signals a decisive call to action for the scholars and intellectuals inside and outside the academy, if a democratic regime of autonomous individuals is to be realized. Reestablishing the translation of the private into the public, as I've previously suggested, requires a necessary rethinking of the notion of the public, its frequent separation from the private, and its alleged independence. Nor can we afford in a neoliberal era the traditional liberal assumption that public interests will be attended to by the state and private interests will be addressed by the individual. Against Margaret Thatcher's insistence that "There is no such thing as society" we need to recall "from exile ideas such as the public good, the good society, equity, justice, and so on— such ideas that make no sense unless cared for and cultivated in the company of others."[61] By reclaiming the best elements of republican and liberal models of the state, of citizenship, and of civic education—taking up the project of autonomy and self-institution as well as a commitment to fundamental rights and freedoms as pedagogical imperatives—it is possible to challenge the disappearance of politics itself. And in this way we can hope to reverse the desperate experiences of fear, uncertainty, and alienation that accompany the painful erosion of individual and social agency.

Defining the university as an *agora*, a space of translation, also demands recognition of a fundamental redefinition of the intellectual vocation. This in turn means rethinking the fashionable dismissal of progressive work in the academy as merely reproducing, through the use of the teacher as ethical exemplar, a form of normative regulation that serves the interests of the dominant order,[62] as well as recent right-wing charges of "anti-Americanism." If, as Bauman has argued, the university has previously functioned to serve the interests of "nation-building, legitimation-seeking powers," the task of intellectuals was to assume the role of custodians and wardens of the rest of the "population-in-need-of-enlightenment-and-cultivation," to serve in the creation, preservation, and continuity of "order" by eliciting obedience through a process of Panoptic regulation.[63] However, the present corporate and military powers, having no use for legitimation, no longer require the services of missionaries, philosophers, or educators. One has only to consider the recent boom in distance education and the appearance of the "online" university to recognize the effective end of "order" as the primary task of power, to see that such education in this instance is about buying and selling skills on/for the open market.

Conclusion

Once we recognize that we can no longer deny our obligations as civic educators through an outdated appeal to social and moral reproduction, the task at hand is to rethink the university as an *agora* and to explore what this space of translation requires. What does it mean to rearticulate private concerns and fears as part of the public interest, as public issues for debate and engagement? What does it mean to demand a different order of "security," emphasizing employment, health care, and education, as opposed to the relentlessly militarized version promoted by our national government? How do we invent a language of community or dare to assert a notion of the public good given the exclusionary legacies of these concepts both nationally and globally? What theoretical tools, what language does this require?

To begin, progressives must confront a deeply troubling contradiction in the relationship between the contemporary academy and society. Within the last two decades, the liberal arts have experienced a boundless proliferation of radical discourses, sub-fields, and disciplines that seek to theorize *the political*, to map social identities, relations, and practices and their implications with power. At the same time, those of us in and out of the university have experienced and continue to experience the disappearance of *politics* itself.[64] Within the university, encroaching corporatization has produced a more hierarchical institutional arrangement, which has translated into the dramatic limitation of academics' influence over the conditions of their work. Within the broader society, we have witnessed the disappearance of public spaces and public access to the means of mass communication—and the decline of meaningful public discussion of issues vital to the health of a democratic society. As my students proved to me beyond doubt, there are alternative Internet spaces for discussions of imminent concern to democracy: breakneck unemployment at a time of record corporate profit that further exacerbates the inequalities between the rich and the poor; the growing immiseration of the poor and communities of color with the dismantling of welfare; a prison industrial complex that houses over two million inmates and the startling realization that the "land of the free" has become the world's biggest jailer; and the ever-expanding militarization of public space and public time. As important as these alternative Internet spaces are, citizens' rights to be informed, which are inevitably tied to access to a free and independent media, are gravely imperiled by an ever-expanding, global corporate media monopoly whose primary interest is not informing the public but maximizing profit. Moreover, a zeitgeist of fear and anxiety—the result of many factors, but topping the list is both the recognition of our collective vulnerability and a real decline in our sense of collective agency—has become widespread, if not terminal.[65]

At first glance, such a tragic state of affairs might be enough to prove once and for all what a growing number of scholars from across the ideological spectrum have long suspected—that the cultural politics of "race, class, and gender" qualifies as neither scholarship nor real politics. I point to this contradiction, however, not to denigrate the advent of those theoretical discourses since the 1960s that have reinvigorated, at least for a time, a flagging liberal arts curriculum. Rather, I'm suggesting that given the ongoing attacks on the humanities that have been given impetus and urgency by the war on terrorism, we must pose difficult questions to theoretical work that justifies itself as "liberatory," questions that address what it is in its own right, the conditions of its own cultural formation, and how it intervenes in the world. On one hand, the apparent inefficacy of new theoretical discourses to revive a waning political culture has little to do with the debates that go on within the humanities and much more to do with events that are largely extra-disciplinary. For example, the increasing corporatization of the university—among the numerous neoliberal political and economic agendas that wreak havoc on public goods and services—has attempted to either assimilate the liberal arts to its own interests or render such disciplines an ornamental, if not entirely irrelevant, endeavor. This has not only perpetuated the decline of funding from the federal government and private institutions, but also has led to the diminished emphasis on liberal arts requirements for undergraduates, the nonrenewal of faculty lines (especially in ethnic and women's studies), and the consolidation or even elimination of whole departments in classics, romance languages, and the social sciences.[66] On the other hand, it may also be observed that many cultural theorists, while engaging and illuminating in some idealist sense, have little to communicate about the changing institutional conditions that threaten their very existence. As Stuart Hall has forcefully argued, "The state didn't send out the secret police to transform higher education into an entrepreneurial sector; we have done that all by ourselves by taking on the ethic of managerialism as the everyday practice of institutional life."[67] Hall's observation demands that progressives rethink how they negotiate their interest in political work in the classroom and in their research, as well as the dictates of a neutral, nonpolitical "professionalism," rejecting the notion that scholarship and commitment are mutually exclusive.[68]

Of course, the increasingly vituperative attacks on the humanities in light of its alleged "multicultural turn" cannot be reduced to the imperatives of corporatization or the aftershocks of the nation's brutal attack. Rather, I want to reiterate that the increasing attacks on the university and other public spaces are part of a broader attack on the regime of political democracy itself. Such assaults are visible in the increasing militarization of everyday life, including, but not limited to, the "patriotic" quashing of dissent in the university and the national news media and the revoking of civil liberties in the name of national security; the relentless privatization and depoliticization of the academy and other public spheres; and the proliferating attacks on difference, which range from the rise of religious fundamentalisms to anti-immigration and zero-tolerance policies to anti–political correctness movements and challenges to affirmative action on college campuses.

Further, with the advent of the war against terrorism, the right has been able to locate an "enemy" as durable as the one it had constructed within national borders (e.g. the so-called "underclass" illegal immigrants, campus radicals . . .). As Ulrick Beck argues:

> Enemy stereotypes therefore delimit democracy in the double sense that they make it possible to put up fences that must call a halt to all self-evident democratic truths and they legitimate the furnishing and delimitation of "democracy-free or low-democracy zones." Here according to established rules, everything can be tried out, planned and implemented which would otherwise be subject to the strictest prohibitions. Examples include: planning and perfecting murder; spending money on horrific weapons systems, the productivity of which "culminates" in their never being used; and many other things . . . . They [enemy stereotypes] empower the powerful and siphon off the consent to do it from the powerless.[69]

The upshot of the efforts in the war on terrorism—now recombined with the war on drugs and crime—is an antidemocratic and militarized state in which citizens lose their rights, freedoms, agency, and security in the name of national safety.

Whatever narrow opening for self-questioning at both the individual and collective level was created in the brief heyday of critical narratives circulating in the academy is now under assault.

In the backlash of jingoistic patriotism, Christian family values and zero-tolerance, self-scrutiny and engaged public dialogue (which are the basis of democracy) are as welcome as a spore of anthrax. Absolutism, certainty, and resolution are the orders of the day. Not surprisingly, given these contexts, a commitment to unmasking where and how the politics of race, gender, class, or nationality have threatened access to and participation in the democratic life of the nation has won progressive scholars a place in what Richard Rorty once infamously referred to as "The Unpatriotic Academy."[70]

Of course, there is some gain beyond intellectual satisfaction in this kind of analysis, in spite of such railing. But at the same time, those of us involved directly in the production and dissemination of knowledge must be ever-critical of our own formations and work. Without question, scholarship over the last three decades has been marked by a vigorous engagement with a variety of radical artistic and intellectual discourses. Such energy is real, and might be reassuring, "if we did not have to remember," Raymond Williams once cautioned, ". . . the comparable liveliness of Weimar culture in the 1920s . . . which, when it came under pressure, was shown to have been all along a double-edged vitality, unified only by its negations, as throughout the whole period of the avant-garde."[71] Then, the target of Williams's animus was a group of intellectuals invested in a theory of ideology that taught despair and political disarmament in the face of Thatcherite revolution. Yet today, intellectual history appears to repeat itself as the same kinds of formalist and (simpler) structuralist analyses of corporate-capitalist and bureaucratic societies like the United States have degenerated into a similar litany of stark negations—of cultural politics and all manner of institutional work as de facto insignificant or in collusion with domination. Such cynicism is visible in two only apparently mutually exclusive gestures. The first is to dismiss the work of critical intellectuals in the university as always already corrupted by their institutional location. The second is to engage oppositional ideas in ways that often reproduce the cult of professionalism and expertise, the historic role of which has been to induce both fear and passivity but also suspicion and anger in professionals in other disciplines and lay peoples. In his history of the

American university, *The Culture of Professionlism*, Burton Bledstein explains:

> The university not only segregated ideas from the public, intellectual segregation occurred with the development of each new department in the university. A department emphasized the unique identity of its subject, its specialized qualities and language, its special distinction as an activity of research and investigation. Any outsider who attempted to pass judgement on an academic discipline . . . was acting presumptuously. In order to further their control over a discipline, professionals particularized and proliferated the possibilities for investigation. . . . The more technical and restricted the individual area of investigation, the more justifiable it became to deny the public's right to know or understand the professional's mission.[72]

My point here is not to denounce complex theoretical work by oppositional intellectuals as such, but to challenge the effectiveness of its relentless insularity in the face of massive antidemocratic assaults on the public sphere.

Nor is my point to deny that the marked tendency of the times is, as Manuel Castells, Zygmunt Bauman, and others have observed, the increasing separation of power from politics, in which extant social institutions appear largely incapable of assistance, much less countering the diminished capacities of democratic public spheres in a global neoliberal era.[73] Bauman's response to these conditions is instructive: it is an insistent demand for more institutional intervention—academic and otherwise—rather than a condemnation of it. As I have already argued, his response to the decline of political agency in the context of globalization is to thwart the processes of militarization, privitization, and corporatization by struggling over remaining public spaces like the university. It seems to me that precisely because of these unprecedented economic and social changes, the pedagogical dimension of all cultural spheres—inside and outside of official educational institutions—remains a crucial site of struggle, though it is a struggle that increasingly reflects the dimensions of David and Goliath.

This suggests that progressive educators should promote, at the university level, genuine education over job training, as well as create linkages with alternative sites outside the university in ways that engage the pedagogical force of the entire culture. Ironically, Raymond Williams anticipated the current state of crisis in university education, and in public life more generally, nearly 35 years ago. In the preface to the revised edition of his *Communications* (1966), Williams accurately predicted the capacities of capital to integrate the social necessity for what he calls "permanent education" for its own purposes:

> The need for permanent education, in our changing society, will be met in one way or another. It is now on the whole being met, though with many valuable exceptions and efforts against the tide, by an integration of this teaching with the priorities and interests of a capitalist society, and of a capitalist society, moreover, which necessarily retains as its central principle (though against powerful pressures, of a democratic kind, from the rest of our social experience) the idea of a few governing, communicating with and teaching the many. . . . Organized economically, in its largest part, around advertising, it is increasingly organized culturally around the values and habits of that version of human personality, human need and human capacity. *This strong and integrated world is capable, I believe, in the coming decades of adapting to its own purposes both politics and education.*[74]

Williams's concept of permanent education has proven extremely useful in rendering visible the pedagogical force of "our whole social and cultural experience," and it explains his lifelong interest not only in the social function of formal educational structures, but also in "what the whole environment, its institutions and relationships, actively and profoundly teaches."[75] But he grasps something else at least as profound. Williams recasts the crisis of contemporary life not simply as a crisis of economics, but as a *crisis of culture*. With uncanny precision, Williams forecasts the capacity of neoliberal social and economic policies to instill in place of democratic principles of autonomy and social justice those "values and habits" that the market puts into place, to invest both literally and affectively in what Cornelius Castoriadis once called the sole grand narrative left—the accumulation of junk and more junk.

Certainly, Williams did not live to see how Francis Fukuyama's "end of history" thesis was a harbinger not only of the decline of communism, but also the effective end of democracy, to the degree that the latter, no longer opposed to the "Evil Empire," became synonymous with the dynamics of capitalism itself. Nor could he have envisioned how the millennial war on terrorism and its "Axis of Evil" have advanced the cause of neoliberal globalization with minimal public outcry (or even awareness). Yet Williams's caution that both education and politics could be subsumed by economic forces seems to anticipate the degree to which societies like the United States have since experienced the relentless privatization and denigration of all things public, the vast consolidation of ownership over media communications, and the vocationalization of higher learning. To paraphrase Williams, we now encounter an imperative for job training in the university that would have sounded crude and embarrassing by late-nineteenth-century standards when something very similar was proposed. Now, again, it is frequently argued that people must be given skills to earn a decent wage within a postindustrial economy to which they must adapt. As *that* syllabus is written, as that mandate for university reform rapidly materializes, there is little room envisioned for intellectuals who engage in critical cultural analysis in the interests of a global public sphere.

Hence it is the task of radical educators to secure not only a space for free inquiry and dissent—especially in times of global crisis—but also the conditions for their own autonomy within the academy. Williams's insights provide an alternative social vision for progressive scholars and cultural workers who are caught up in the contradictions between a neoliberal push to depoliticize and corporatize formal educational spaces and a neoconservative pull to use such spaces to reregulate and publicize the most private, intimate concerns of citizens, particularly around issues relating to the body and sexuality. In contrast to a form of permanent education created by the interests of capital, he proposes an educational vision rooted in the interests of political agency and popular democratic governance:

> Against this kind of permanent education, already well organized and visibly extending its methods and its range, an integrated

alternative is now profoundly necessary. I have seen something of the plans, in many countries, for a permanent education of a democratic and popular kind: programmes for family care, for the improvement and extension of schools, universities and further education, for the public safeguarding of natural beauty, for the planning of towns and cities around the needs of leisure and learning, for the recovery of control and meaning in work. It is in the spirit of this kind of programme that I discuss communications, the field in which one or other version of a permanent education will be decisive.[76]

There is no question, as Williams makes clear, that one form of permanent education "will be decisive." And it is equally apparent that the prospects for instituting a genuinely democratic and popular educational program within the university and all educational institutions—and hence the promise of substantive democracy for future generations—are in doubt.

My intention has not been to isolate education as the key to revitalizing a waning political democracy. The assumption that education alone can alter iniquitous social conditions is both unrealistic and naïve in that it denies how such institutions are affected by those political and economic conditions that they are allegedly supposed to counter. Rather than offering education as a kind of panacea for contemporary social problems, it seems to me necessary nonetheless in the current state of seige that progressive educators and others address what role the university might play as part of a broader effort to secure *the very conditions* for democratic participation in public life. In the interests of real homeland security, we have to open up rather than close down our classrooms to dialogue and debate over those contemporary issues and hot-button topics that most concern our students, as we attempt to open up access to venues for broad public conversation and civic participation.

Chapter 2

Academic Culture, Intellectual Courage, and the Crisis of Politics in an Era of Permanent War

The question that I would like to raise is this: Can intellectuals, and especially scholars, intervene in the political sphere? Must intellectuals partake in political debates as such, and if so, under what conditions can they interject themselves efficiently? What role can researchers play in the various social movements, at the national level and especially at the international level—that is, at the level where the fate of individuals and societies is increasingly being decided today? Can intellectuals contribute to inventing a new manner of doing politics fit for the novel dilemmas and threats of our age?

—Pierre Bourdieu[1]

Introduction

Pierre Bourdieu, a French Sociologist, was deeply concerned about the role that academics might play as a progressive force in politics. He believed that academics were indispensable, given their rigor as researchers, writers, and teachers in creating the pedagogical conditions that both furthered social and economic justice and challenged the forms of symbolic and material domination being exercised globally, especially under neoliberalism. Rejecting the commonplace assumption that academic work should be separate from the operations of politics, he reclaimed the role of the intellectual as an engaged social agent and "maintained that intellectuals have a fearsome form of social responsibility."[2] Following Edward Said, Noam Chomsky, Howard Zinn, and others, Bourdieu argued that for academics to become engaged intellectuals they had to repudiate the cult of professionalism that has often positioned educators as narrow specialists, unencumbered by matters of ethics, power, and ideology, and wedded to a sterile objectivity that largely serves to justify a retreat into a world of banal academic rituals and unapologetic escapism. Against the cult of professionalism, Bourdieu posited the notion of committed intellectuals in search of "realist" utopias. But committed scholarship, for Bourdieu, does not mean limiting politics, pedagogy, or social change to the world of texts or the narrow province of discourse. Nor does committed scholarship and pedagogy provide an excuse for those intellectuals who often "mistake revolutions of the order of words, or texts, for revolutions in the order of things, to mistake verbal sparring at academic conferences for interventions in the affairs of [public life]."[3] Nor does a gesture toward engaged scholarship warrant its translation into sound bytes for trendy media intellectuals. Bourdieu dismissed such intellectuals as second-rate "scholars of the obvious" who shamelessly offer up 30-second commentaries on just about anything and everything, all of which is generally couched in glib political slogans that serve to misinform and naturalize dominant discourses while reproducing existing relations of power.[4]

According to Bourdieu, academics had to not only engage in a permanent critique of the abuses of authority in the media and society, but also address the deadening scholasticism that often characterized work in the academy. This was not simply a call for them to renounce an all too common form of political irrelevance rooted in the mantra of professional objectivity, neutrality, and distance that inveighed against connecting higher education to the public realm and scholarship to larger social issues, but also an attempt to convince intellectuals that their own participation in the public realm should never take place at the expense of their artistic, intellectually rigorous, or theoretical skills. The meaning of what it meant to be a public intellectual could not serve as an excuse for either a narrow scholasticism, anti-intellectualism, or withdrawal into academic irrelevance that willingly separate commitment from scholarship. Nor is assuming the role of an intellectual for the public an excuse to substitute a celebrity-like, public-relations posturing for the important work of providing alternative, rigorous analyses and engaging important social issues through individual and collective struggles. Bourdieu wanted intellectuals—academic and non-academic alike—to organize and become a collective force for fighting against a range of injustices, preserving the benefits of the welfare state, and transforming the neoliberal state into an inclusive democracy. This becomes clear in his call to intellectuals to get actively involved on a permanent basis in the most important struggles of their time:

> When it comes to restructuring public services, intellectuals, writers, artists, scientists and others have to play a decisive role. First of all they can help break the monopoly of the technocratic orthodoxy in the media. But furthermore they should work on an organized and permanent level—and not just in times of crisis at occasional meetings—with those capable of giving society's future a direction.[5]

We believe that academics, because of their freedom and privileges, have a particularly important role to play as engaged public intellectuals at this particular moment in history. One of the most dangerous problems they now face is a repressive right-wing government that is massively redistributing wealth from the poor to wealthy individuals and powerful corporations, undermining civil liberties, and promoting military aggression abroad. Another threat to democracy, critical intellectuals, and citizens alike comes from the neoconservative attack on affirmative action in education, immigration, the welfare state, and those secular values separating church and state. In addition, there is the danger posed by a virulent neoliberalism, with its all-consuming emphasis on market relations, commercialization, privatization, and the creation of a worldwide economy of part-time workers. In what follows, we want to take

Bourdieu's emphasis on the role of academics as engaged intellectuals seriously; but rather than simply defend that role, we want to explore how it might be made more concrete by situating it within a mode of analysis and public witnessing that addresses a number of important political and educational issues, such as militarism, the attack on the welfare state, and the war against youth.

Permanent War at Home and Abroad

Where did this idea come from that everybody deserves free education? Free medical care? Free whatever? It comes from Moscow. From Russia. It comes straight out of the pit of hell.[6]

My goal is to cut government in half in twenty-five years, to get it down to the size where we can drown it in the bathtub.[7]

We are living in dangerous times in which a new type of society is emerging unlike anything we have seen in the past—a society in which symbolic capital and political power reinforce each other through a media apparatus largely controlled by ten major corporations, which have become a cheerleading section for dominant elites and corporate ruling interests. This is a society increasingly marked by an attack on democracy, a poverty of critical public discourse, and, as Debbie Riddle and Grover Norquist in the two quotes above suggest, a virulent contempt for social needs and the public good. It is also a social order that seems incapable of questioning itself, even as it wages war against the poor, youth, women, people of color, and the elderly at home, and elevates the doctrine of "permanent war" as the basis for an aggressive and arrogant policy abroad. Unprecedented military might combines with an arrogance of power that allows the White House to redefine the philosophy of international politics by proclaiming that U.S. military power should remain beyond challenge. Hence, America has the right to launch preemptive strikes against perceived threats, and the United States can conduct its foreign policy unilaterally, without the need for international support. At home and abroad, diplomacy and dialogue have been replaced by a political culture in Washington, D.C. that devours democratic citizenship, shreds the social contract, and criminalizes social problems. Under the shadow of war and a ruthless neoliberal assault on all things public, a society is emerging in the United States that makes it increasingly difficult for young people and adults to appropriate a critical language outside of the market that would allow them to analyze private problems as public concerns or to relate public issues to private considerations.

The recent war with Iraq will change the way the United States relates to the rest of the world as well as how it addresses the most pressing problems Americans face in their everyday lives. But the danger we face as a nation—a danger that needs to be engaged, in part, by university academics—is not only related to the war in Iraq; it is also related to the silent war at home, especially since the Iraq war and the war against terrorism are being financed from cuts in domestic funding on health care, children's education, and other public services. It would be a tragic mistake for educational critics either to separate the war in Iraq from the many problems Americans face at home or to fail to recognize how war is being waged by this government on multiple fronts.

The war against terrorism is part and parcel of the war against democracy at home. Slavoj Žižek claims that the "true target of the 'war on terror' is American society itself—the disciplining of its emancipatory excesses."[8] George Steinmets argues that the current state of emergency represents a new shift in the mode of political power and regulation. He claims that:

The refocusing of political power on the level of the American national state has been most evident in the area of U.S. geopolitical strategy (unilateralism and preemptive military strikes), but much of the new regulatory activity has focused on the state apparatus itself and the 'domestic' level of politics, with the creation of a huge new government agency (the department of Homeland Security), transformations of the legal system (e.g., secret trials and arrests, indefinite detentions), and intensified domestic surveillance: first with the 2001 USA PATRIOT Act, which dramatically relaxed restrictions on search and seizure; then with the [now defunct] Total Information Awareness program, which collects and analyzes vast amounts of data on private communications and commercial transactions; and most recently with the proposed domestic Security Enhancement Act of 2003.[9]

Both observations are partly right. The Bush "permanent war doctrine" is not just aimed at alleged terrorists or the excesses of democracy, but also against disposable populations in the homeland, whether they be young black men who inhabit our nation's poorest neighborhoods or those unemployed workers who have been abandoned by the flight of capital as well as by government. The financing of the war in Iraq is buttressed by what Vice President Dick Cheney calls the concept of "never ending war." This is a concept that declares permanent war as a continuous state of emergency and brings into play a fundamentally new mode of politics. In a commencement speech given recently at the United States Military Academy, Cheney provides a succinct outline of the permanent war concept:

> The battle of Iraq was a major victory in the war on terror, but the war itself is far from over. We cannot allow ourselves to grow complacent. We cannot forget that the terrorists remain determined to kill as many Americans as possible, both abroad and here at home, and they are still seeking weapons of mass destruction to use against us. With such an enemy, no peace treaty is possible; no policy of containment or deterrence will prove effective. The only way to deal with this threat is to destroy it, completely and utterly.[10]

The apocalyptic tone of his comments do more than cover up the fictive relationship between Iraq and 9/11; it also serves to legitimate a bloated and obscene military budget as well as economic and tax policies that are financially bankrupting the states' budgets, destroying public education, and plundering public services. The U.S. government plans to spend up to $400 billion to finance the Iraqi invasion and the ongoing occupation, while it allocates only $16 billion to welfare programs, which cannot possibly address the needs of over 34.8 million people who live below the poverty line, many of them children, or the 43 million without health insurance, or the millions now unemployed because of diminished public services and state resources. Even more remarkable is the recent report by United Press International that at Fort Stewart, Georgia, over 600 "sick or injured members of the Army Reserves and National Guard are warehoused in rows of spare, steamy and dark cement barracks in a sandy field waiting [for weeks] for doctors to treat their wounds or illnesses." The Bush administration spends millions to equip these young men with the weapons of war, but shortchanges them on decent medical care when they return home wounded and sick. This government ships young men and women to Iraq and Afghanistan in record time, but once they return to the U.S. in need of medical treatment, they often have to wait for weeks, if not months. While $723 billion dollars are allocated for tax cuts for the rich, state governments are cutting a total of $75 billion in health care, welfare benefits, and education. Many states such as California, with a $38.8 billion-dollar budget gap, are implementing "drastic cuts in public school services and the withholding of potentially life saving medicine from seriously ill patients."[11] Oregon has cut back on the school year and cut out extracurricular activities. Many states are laying off crucially needed police, fire, and health workers. Massachusetts has laid off hundreds of sorely needed public school teachers. Many states such as New York and Pennsylvania have cut back on crucial basic programs such as medical services to the seriously ill, programs for the disabled, and prescription drug benefits for the poor.

The sheer inhumanity the Bush administration displays toward the working poor and children living near the poverty level can be seen in the decision by Republicans in Congress to eliminate from the 2003 tax bill the $400 child credit for families with incomes between $10,000 and $26,000. The money saved by this cutback will be used to pay for the reduction on dividend taxes. As a result, as Bill Moyers observes, "[Twelve] million children are punished for being poor, even as the rich are rewarded for being rich."[12] Bush's tax cuts are an insult to any viable notion of social justice. For instance, according to the Citizens for Tax Justice, "the top 1 percent of Americans—those making over $335,000 a year or more—will get, on the average, tax breaks worth $103,899," while the bottom "20 percent—those making under $16,000—will get all of $45 over the four-year period."[13]

As engaged intellectuals, academics and students need to connect these multiple attacks on the poor and much needed public services to an expanded political and social vision that refuses the cynicism and sense of powerlessness that accompanies the destruction of social goods, the corporatization of the media, the dismantling of workers' rights, and the appropriation of intellectuals. Against this totalitarian onslaught, educators need to develop a language of critique and possibility, one that connects diverse struggles, uses theory as a resource, and defines politics not merely as critical but also as a transformative intervention into public life. We need a language that relates the discourse of war to an attack on democracy at home and abroad, and we need to use that language in a way that captures the needs, desires, histories, and experiences that shape people's daily lives. Similarly, as democratic institutions are downsized and public goods are offered up for corporate plunder, those educators who take seriously the related issues of equality, human rights, justice, and freedom face the crucial challenge of formulating a notion of the political suitable for addressing the urgent problems now facing the twenty-first century—a politics that as Zygmunt Bauman argues, "never stops criticizing the level of justice already achieved and seeking more justice and better justice."[14]

As the federal government is restructured as a result of right-wing assaults by the Bush administration, it has dramatically shifted its allegiance away from providing for people's welfare, protecting the environment, and expanding the realm of public good. The social contract that addresses matters of injury and protection, allocates social provisions against life's hazards, and lies at the heart of a substantive democracy has been nullified. As the social contract is shredded by Bush's army of neoliberal evangelicals, neoconservative hardliners, and religious fundamentalists, government relies more heavily on its militarizing functions, giving free reign to the principle of national security at the expense of public service and endorsing property rights over human rights. A spreading culture of fear in an age of automated surveillance and repressive legislation has created a security state that gives people the false choice between being safe or free. As a result, constitutional freedoms and civil liberties are compromised as the FBI is given the power to sieze the records of library users and bookstore customers, the CIA and Pentagon are allowed to engage in domestic intelligence work, and the PATRIOT Act allows people to be detained indefinitely in secret without access to either lawyers or family members. Under such circumstances, as Arundhati Roy argues, "the fundamental governing principles of democracy are not just being subverted but deliberately sabotaged. This kind of democracy is the problem, not the solution." The shadow of authoritarianism becomes increasingly darker as society is organized relentlessly around a culture of fear, cynicism, and unbridled self-interest—a society in which the government promotes legislation urging neighbors to spy on each other and the President of the United States endorses a notion of patriotism based on moral absolutes and an alleged mandate to govern, which comes directly from God (with a little help, of course, from Jeb Bush and the U.S. Supreme Court). According to the author of *The Faith of George W. Bush*, Stephen Mansfield, Bush told a fellow Texan evangelist prior to his presidential candidacy announcement: "I feel like God wants me to run for President. I can't explain it, but I sense my country is going to need me. . . . I know it won't be easy on me or my family, but God wants me to do it."[15]

Increasingly, we are told by President Bush, John Ashcroft, Dick Cheney, and others that patriotism now legitimates unaccountable power and unquestioned authority, defined rather crudely in the dictum "Either you are with us or with the terrorists."[16] Such absolutes, of course, have little respect for difference, dissent, or, for that matter, democracy itself. This new politics depends less on public engagement, dialogue, and democratic governance than with a heavy reliance on institutions that rule through fear and, if necessary, brute force. The devaluation of politics and the depoliticization of public engagement in the United States has taken an ominous turn with the ongoing militarization of language, public space, and everyday life. Communities are now organized around fear rather than civic courage, compassion, or democratic values. Increasingly, power is being used by the Bush administration to promote what Sheldon Wolin calls, "empire abroad and corporate power at home,"[17] increasingly mediated and legitimized through the rhetoric of war, fear, and antiterrorism.

As Ulrich Beck has argued, the language of war has taken a distinctly different turn in the new millennium.[18] War no longer needs to be ratified by Congress since it is now waged by various government agencies that escape the need for official approval. War has become a permanent condition adopted by a nation-state that is largely defined by its repressive functions in response to its powerlessness to regulate corporate abuses, provide social investments for the populace, and guarantee a measure of social freedom. The concept of war occupies a strange place in the current lexicon of foreign and domestic policy. It no longer simply refers to a war waged against a sovereign state such as Iraq, nor is it merely a political referent for engaging in acts of national self-defense. The concept of war has been both expanded and inverted. It has been expanded in that it has become one of the most powerful concepts for understanding and structuring political culture, public space, and everyday life. Wars are now waged against crime, drugs, terrorism, even obesity, among a host of alleged public disorders. Wars are not declared against foreign enemies but against alleged domestic threats. The concept of war has also been inverted in that its mechanisms for legitimation no longer invoke the concept of social justice—a relationship that emerged under President Lyndon Johnson and was exemplified in the war on poverty. War has become a metaphor for legitimating a zone of power in which the national security state displaces its more democratic role (social, egalitarian, peaceful, and democratic). As Susan Buck-Morss observes:

The U.S. national security state is a war machine positioned within a geopolitical landscape. It must have a localizable enemy for its power to appear legitimate; its biggest threat is that the enemy disappears. But given a war, even a Cold War, and now given an ill-defined yet total "war on terrorism," the declared "state of emergency" is justification for suspending the rights and freedoms of citizens. It justifies arresting and holding individuals without due process. It justifies killing and bombing without oversight or accountability. It justifies secrecy, censorship, and a monopoly over the accumulation and dissemination of information. All of these state practices are totalitarian, of course.[19]

Under the reign of the national security state, war is now defined almost exclusively as a punitive and militaristic process. This can be seen in the ways in which social policies are now criminalized so that the war on poverty is now a war against the poor, the war on drugs is now a war waged largely against youth of color, and the war against terrorism is now a war against immigrants, domestic freedoms, and dissent itself. In the Bush, Perle, Rumsfeld, and Ashcroft view of terrorism, war is individualized as every citizen becomes a potential terrorist who has to prove that he or she is not dangerous. Under the rubric of emergency time, which feeds off government-induced media panics, war provides the moral imperative to collapse the "boundaries between innocent and guilty, between suspects and non-suspects."[20] War provides the primary rhetorical tool for articulating a notion of society as a community organized around shared fears rather than shared responsibilities and civic courage. War is now transformed into a slick, Hollywood spectacle designed to both glamorize a notion of hyper-masculinity fashioned in the oil fields of Texas and fill public space with celebrations of ritualized militaristic posturing, touting the virtues of either becoming part of "an Army of one" or indulging in commodified patriotism by purchasing a new Hummer. Of course, this corrupt version of patriotism excludes certain class and racial minorities who cannot buy their participation in it.

War as a spectacle combines easily with the culture of fear to divert public attention away from domestic problems, define patriotism as consensus, and further the growth of a police state. The latter can be seen not only in the passage of the PATRIOT Act and the suspension of civil liberties, but is also evident in the elimination of those laws that traditionally separated the military from domestic law enforcement and offered individuals a vestige of civil liberties and freedoms. The political implications of the expanded and inverted use of war in its metaphorial and actual manifestations is also evident in the war against "big government," which is really a war against the welfare state and the social contract itself—that is, a war against the notion that everyone should have access to decent education, health care, employment, and other public services. Of course, while the war against big government has been relegated to a knee-jerk political slogan by Republicans and conservative Democrats alike, one has to search far and wide to hear a peep out of this group about the threat that big corporations pose to democracy or, for that matter, an articulation of what the role of responsible government might actually be. One of the most serious issues to be addressed in the debate about Bush's concept of permanent war is the effect it is having on one of our most vulnerable populations, children, and the political opportunity this issue represents for constructing a language of both opposition and possibility.

The War Against Children and the Politics of Neoliberalism

Wars are almost always legitimated in order to make the world safe for "our children's future," but this rhetoric belies how their future is often denied by the acts of aggression put into place by a range of ideological state apparatuses. This would include the horrible effects of the militarization of schools, the use of the criminal justice system to redefine social issues such as poverty and homelessness as violations of the social order, and the subsequent rise of a prison-industrial complex as a way to contain disposable populations such as youth of color who are poor and marginalized. Under the rubric of war, security, and antiterrorism, children are "disappeared" from the most basic social spheres that provide the conditions for a sense of agency and possibility, as they are rhetorically excised from any discourse about the future. The "disappearing" of children is made more concrete and reprehensible with the recent revelation that three children between the ages of 13 and 15 are being held without legal representation as enemy combatants in possibly inhumane conditions at the military's infamous Camp Delta at Guantanamo Bay, Cuba. One wonders how the Bush administration reconciles its construction of a U.S. gulag for children with their fervent support of family values and the ideology of "compassionate conservativism."

The Bush administration's aggressive attempts to reduce the essence of democracy to profitmaking, shred the social contract, elevate property rights over human rights, make public schools dysfunctional, and promote tax cuts that will limit the growth of social programs and public investments fail completely as public policy when applied to the vast majority of citizens, but especially so when they are applied to children. And yet, children provide one of the most important ethical and political referents for exposing and combating such policies. President Bush and his administrative cronies are redefining the nature of the social contract so as to remove it from the realm of politics and democracy. In doing so, Bush and his followers are not only consolidating their political power; they are also pushing through harsh policies and regressive measures. Such policies end up cutting basic services and public assistance for the poor, the elderly, and the infirmed, sacrificing American democracy and individual autonomy for the promise of domestic security, and allocating resources and tax breaks to the rich through bailouts and regressive tax cuts. Bush's rhetoric of "permanent war" and antiterrorism has done more than create a culture of fear and a flood of jingoistic patriotism; it has also covered up those neoliberal tax polices for the rich that are part of the war waged against public goods, the very notion of society, and those marginalized by class and race. As Jeff Madrick observes:

> Narrow politics, of course, can partly account for the Bush administration's tax proposals. The tax cuts disproportionately benefit the wealthy, which, after all, is Bush's natural political constituency. But Bush's policies may, in fact, be explained by another, more radical agenda. Extensive tax cuts will require Congress to limit the growth of social programs and public investment and undermine other programs altogether. If that is your vision of the best direction America can take, the strategy makes some sense. So, we were wrong about how dividend tax cuts stimulate growth, you can almost hear the Bush advisers thinking. No problem. Rising deficits will inevitably force Congress to starve those "wasteful" social programs. The prospective high deficits may even make it imperative to privatize Social Security and Medicare eventually. Social spending is the problem, goes the argument, not tax cuts.[21]

Starving social programs and destroying public institutions have their most immediate effects on children, especially those who are poor, lack adequate resources, and are trapped in a cycle of poor health and structures of inequality. Making visible the suffering and oppression of young people cannot help but challenge the key assumptions of "permanent war" policies designed to destroy public institutions and prevent government from providing important services that ameliorate ignorance, poverty, racism, inequality, and disease. The well-being and future of youth offer a crucial rationale for engaging in a critical discussion about the long-term consequences of current administration policies, especially those driven by neoliberalism, an issue we address below in more detail.

As society is defined through the culture, values, and relations of neoliberalism, the relationship between critical education, public morality, and civic responsibility as a condition for creating thoughtful and engaged citizens is sacrificed all too willingly to the interest of finance capital, corporate greed, and the logic of profitmaking. Under the auspices of neoliberalism, citizens lose their public voice as market liberties replace civic freedoms and society increasingly depends on "consumers to do the work of citizens."[22] Similarly, as corporate culture extends even deeper into the basic institutions of civil and political society, there is a simultaneous diminishing of non-commodified public spheres—those institutions such as public schools, churches, noncommercial public broadcasting, libraries, trade unions, and various voluntary institutions engaged in dialogue, debate, and learning—that address the relationship of the self to public life and social responsibility to the broader demands of citizenship, and also provide a robust vehicle for public participation and democratic citizenship. As Edward Herman and Robert McChesney observe, such non-commodified public spheres have played an invaluable role historically "as places and forums where issues of importance to a political community are discussed and debated, and where information is presented that is essential to citizen participation in community life."[23] Without these critical public spheres, corporate power often goes unchecked and politics becomes dull, cynical, and oppressive.[24] More important, in the absence of such public spheres it becomes more difficult for citizens to challenge the neoliberal myth that citizens are merely consumers and that "wholly unregulated markets are the sole means by which we can produce and distribute everything we care about, from durable goods to spiritual values, from capital development to social justice, from profitability to sustainable environments, from private wealth to essential commonweal."[25] As democratic values give way to commercial values, intellectual ambitions are often reduced to an instrument of the entrepreneurial self, and social visions are dismissed as hopelessly out of date. Public space is portrayed exclusively as an investment opportunity, and the notion of the public increasingly becomes a metaphor for disorder, that is, the public becomes synonymous with disrepair, danger, and risk, as in, for example, public schools, public transportation, public restrooms, public parks, etc. Anyone who does not believe that rapacious capitalism is the only road to freedom and the good life is dismissed as either a crank or worse. Hence it is not surprising to read an editorial in *The Economist*, in which youthful critics of capitalism are dismissed with the nasty claim that "Dwelling too long on their bogus concerns is apt to rot the intellect."[26]

In the absence of open public spaces, it has become much easier for advocates of neoliberalism to eliminate the most basic social provisions of the welfare state, weaken the power of unions, enhance the influence of corporate power over all aspects of daily life, wage war on the environment, leave citizens isolated and disarmed in the face of a worldwide culture of insecurity and fear, and wage class and racial warfare against the poor, immigrants, and people of color. Academics must address all of these issues as part of a pedagogy of responsibility and a politics of commitment. But what is most alarming as a result of the spread of neoliberalism, as we have mentioned earlier in this chapter, is the way in which the social contract that connects adult responsibility to the welfare of youth and a belief in the future has been ruptured. Traditionally, the liberal, democratic social contract has been organized around a commitment to fairness, justice, generosity, and an insistence that government plays a vital part in providing an infrastructure of support and security with respect to health care, housing, and education, as well as those basic services that address both the opportunities and the hazards in people's lives. The social contract recognizes the potential for human injury and provides the conditions for a safe, healthy, and educated life as well as crucial safeguards against sickness, growing old, and unemployment. Not only is the social contract foundational for any viable and inclusive democracy but also, and most important, for providing a decent future for generations of young people.

With the election of George W. Bush as the president of the United States, the forces of neoliberalism have become more intensified as the social contract has been revoked and youth are abandoned with the collapse of democratic values and political, ethical, and social concerns. On almost every political, economic, cultural, and educational front, the market forces of privatization, deregulation, and capital are radically altering the national and global landscape, and the effect on young people has been devastating.

Youth now constitute a "crisis," which has less to do with improving the future than with denying it. This lack of concern for the health, rights, and quality of children's lives reflects the ideology that underlies neoliberal capitalism and its various expressions both at home and globally. The devaluation of children runs through various government policies that have shaped the last two decades. Increasingly, children are subject to conditions in the larger social order that reveal not only the social Darwinian impulse of a society that wants to abandon anyone who is not viable economically (either as a producer or consumer) and consigns the less technologically adept to low-wage, unskilled work. The degree to which American society has lapsed into a kind of barbarism can be measured by the growing refusal to pay attention to the needs of its children. The contradiction is most evident in the ongoing suppression of children's rights, the repression of their voices, and their growing perception either as a threat or as simply disposable in the neoliberal equation. Children are viewed as unfit to be free agents and utterly infantalized, reduced to complete dependence on adults.[27] Yet, when adult society wants to punish children, they are treated as adults and subject to the most brutal machinations of the criminal justice system, including incarceration in adult prisons and the death penalty. Child killing has become so integrated into public policy that there was barely a whimper of protest when Cruz Bustamante, a leading Democratic candidate no less in the 2003 California race for governor, suggested that he would "cast a vote with a tear in my eye to execute 'hardened criminals' as young as 13."[28]

Youth are removed from the moral concerns of society because they are viewed as disposable and unproductive. Their fate is not unlike that of the new poor, who under the reign of neoliberalism are banished from visibility as they are removed from the discourse of social investments and viewed largely through the language of containment. Zygmunt Bauman's comments about the poor extend to those youth in whom society has chosen not to invest. He observes:

> While banishing the poor from the streets, one can also banish them from the community of humans, from the world of ethical duty. This is done by rewriting the story from the language of deprivation to that of depravity. The poor supply the "usual suspects" rounded up to the accompaniment of public hue and cry whenever a fault in the habitual order is detected. The poor are portrayed as lax, sinful and devoid of moral standards. The media cheerfully cooperate with the police in presenting the sensation-greedy public lurid pictures of the crime-, drug- and sexual promiscuity-infested "criminal elements" who find their shelter in the darkness of mean streets.[29]

Some academics are once again engaging with politics though they are not rallying around youth as a referent for thinking about the future. The future and its relationship to democracy has become a matter of great urgency for progressive academics because of the intensified threat posed by the Bush administration to civil liberties; his unilateral aggression against Iraq; his dangerous display of arrogance around the globe, and his reckless handling of the economy. Unfortunately, while a few academics have entered into a public debate around the role of the United States in these areas, they have had almost nothing to say about how these issues affect young people. Youth rarely figure in public conversations about the curtailing of civil rights and liberties, the dismantling of big government services, the rise of the security state, and the profound changes that are driving globalization and U.S. imperialism.

The contributions of Ed Herman, Noam Chomsky, Senator Robert Byrd, Marian Wright Edelman, and a few others notwithstanding, there is almost no mention of children in debates about empire, war, and foreign policy. For example, the moral quality of U.S. foreign policy is rarely invoked in reference to the enormous suffering and deaths it has imposed on the children of Iraq as a result of the U.S. bombing in 1991 and the sanctions imposed after the war. During the 1991 war, Iraq lost a substantial part of its electrical grid, which powered equipment in its water and sewage plants. Of the 20 electric generating plants, over 17 were either damaged or completely destroyed. One consequence was the breakdown of water, sewage, and hospital services and the spread of various water-borne diseases such as dysentery. Anupama Roa Singh, one of the country directors for the United Nation's Children's Fund (UNICEF), has claimed that over half a million Iraqi children under the age of 5 have died since the imposition of UN sanctions over a decade ago. The BBC reported in 1998 that 4,000 to 5,000 children in Iraq were dying every month from treatable diseases that spread because of bad diets and the aforementioned breakdown of the public infrastructure. Against this murderous reality, it becomes more difficult to mount a convincing humanitarian argument for the current U.S. intervention in Iraq—not only because it's clear that the murder and suffering of the children of Iraq will be intensified as a result of the war and the occupation, but also because it undercuts any moral discourse that the United States uses to defend its efforts to "liberate" and "rebuild" Iraq. In an impassioned speech before the U.S. Senate, Senator Robert Byrd raised this issue with incisive clarity. He perceptively noted, "I must truly question the judgment of any President who can say that a massive unprovoked military attack on a nation which is over 50 percent children (under the age of 15) is 'in the highest moral traditions of our country.'"[30] Bush's talk about the moral and democratic imperative to promote regime change, eliminate the axis of evil, and bring freedom to Iraq (and any other country the U.S. opposes) strikes a cruel and hypocritical note in light of the role the United States has played in the death of over half a million children in Iraq. And is it not the same population—the people the Bush administration wants to free—who will pay the ultimate price for the current invasion and occupation by the United States? A recent study, "The Impact of War on Iraq Children," published before the U.S. occupation of Iraq, claimed that children under 18—13 million in all—are "at a grave risk of starvation, disease, death and psychological trauma," and that they are worse off now than they were just before the outbreak of war in 1991.

According to a study published by UNICEF, the recent U.S. invasion "has worsened the health hazards, disrupting clean water supplies, damaging sewage systems and halting rubbish collections."[31] During the recent war, 8,000 civilians were killed along with over 30,000 troops, most of whom were conscripted teenagers. In the aftermath of the war, as the respected journalist John Pilger observes, the "biggest military machine on earth, said to be spending up to $5 billion-a-month on its occupation of Iraq, apparently cannot find the resources and manpower to bring generators to a people enduring [punishingly high] temperatures . . . almost half of them children, of whom eight percent, says UNICEF, are suffering extreme malnutrition."[32]

Astonishingly, government officials are willing to not only ignore the suffering that war brings to the most vulnerable populations, but also defend the slaughter of children as politically expedient. For instance, Madeleine Albright, former U.S. Secretary of State under Bill Clinton, appeared on the news program *60 Minutes* on May 12, 1996, and was asked the following question by the show's host, Leslie Stahl: "We have heard that a half a million children have died [because of sanctions against Iraq]. I mean that's more children than died in Hiroshima. And—you know, is the price worth it?" Albright responded: "I think this is a very hard choice, but the price—we think the price is worth it." How might the parents of Iraqi children feel about this type of cruel political expediency? Does regime change mean that Iraqi civilians, especially children, should be targeted as part of a military and political strategy? Does the liberation of Iraq by Bush and the "neocon" warriors justify dropping cluster bombs and uranium-tipped shells on defenseless populations or, for that matter, securing oil fields but allowing the wholesale looting of Iraq's national treasures, while doing very little to provide basic services such as electricity and fresh water?

The moral insensitivity and inhumanity that underlie U.S. policy toward Iraq cannot be reduced simply to the expediency of its antiterrorism campaign or its need to seize Iraq's rich oil reserves. The roots of this indifference to the rights and needs of children, if not human life in general, must be understood within the larger framework of neoliberalism. As both an economic policy and political strategy, neoliberalism refuses to sustain the social wage, destroys those institutions that maintain social provisions, privatizes all institutions associated with the public good, and narrows the role of the state to both a gatekeeper for capital and a policing force for maintaining social order and racial control. As we mentioned earlier, as an economic policy, neoliberalism allows a handful of private interests to control all aspects of society, and defines society exclusively through the privileging of market relations, deregulation, privatization, and consumerism. As a political philosophy, neoliberalism construes a rationale for a handful of private interests to control as much of social life as possible to maximize their financial investments. Unrestricted by legislation or government regulation, market relations as they define the economy are viewed as a paradigm for democracy itself. Central to neoliberal philosophy is the claim that the development of all aspects of society should be left to the wisdom of the market. Similarly, neoliberal warriors argue that democratic values be subordinated to economic considerations, social issues be translated as private manifestations of moral character, part-time labor replace full-time work, trade unions be weakened, and everybody be treated as a customer. Within this market-driven perspective, the accumulation of capital takes precedence over social justice, the shaping of socially responsible citizens, and the building of democratic communities. Neoliberalism not only separates politics from economic power, destroys the public sector, and transforms everything according to the mandates of the market; it also obliterates public concerns and cancels out the democratic impulses and practices of civil society by either devaluing or absorbing them within the logic of the market. This is what Milton Friedman, the reigning guru of neoliberalism, means in *Capitalism and Freedom* when he argues that "The basic problem of social organization is how to co-ordinate the economic activities of large numbers of people."[33] There is no language here for recognizing antidemocratic forms of power, developing non-market values, or fighting against injustice in a society founded on deep inequalities. Hence, it is not surprising that Friedman can argue without irony that he does not "believe in freedom for madmen or children."[34] Nor should it be surprising that under neoliberalism children are considered valuable only in the most reductive economic terms.

More recently, the debate about neoliberalism has been linked to the narrowing of public debate brought on, in part, by the concentration of power in the hands of the relatively few corporations that now control the media. Monopoly capital is increasingly linked to not only inequality, the war against big government, and the abrogation of the social contract, but also to the weakening of civil liberties and basic freedoms. Yet, once again, little has been said by academics and journalists about how neoliberalism has impacted youth and how neoliberal policies are related to the ongoing war against young people. While the official discourse about highjacking civic freedoms drapes itself in the mantle of national security, secrecy, and patriotism, the repressive policies that underlie the rhetoric have been alive and well long before the terrorist attacks on September 11. The short list includes the Palmer raids of the 1920s, the internment of Japanese Americans during the Second World War, the McCarthy hysteria of the 1950s, and the illegal FBI domestic counterintelligence program (COINTELPRO), conducted between 1956 and 1971, whose sole purpose was to "neutralize" politically dissident groups such as the antiwar and civil rights movements and the Black Panthers. These are powerful examples of how repression has rarely been on the side of either security or justice; but what must be added to this often cited historical record are the various modes of repression that youth have been experiencing since the 1980s.

As the social contract between adult society and children disappears, the old ideology that saw youth as an investment and source of democratic renewal has given way to pure repression. For instance, children, especially youth of color, are increasingly portrayed as a danger to society and an element to be monitored and contained. The consequences of such views in social policies can be seen in the intensified application of profiling, especially among urban youth, the widespread use of random drug testing of public school students, physical searches, and the increased presence of police and the application of zero-tolerance laws in the schools. As the state is increasingly reconfigured as a conduit for the criminal justice system, it withdraws from its liberal role of investing in the social, and now punishes those young people who are caught in the downward spiral of its economic policies. Punishment, incarceration, and surveillance have come to represent the role of the new state. One consequence is that the implied contract between the state and citizens is broken, and social guarantees for youth as well as civic obligations to the future vanish from the public agenda. Crucial issues such as homelessness, poverty, and illiteracy among youth are now viewed as individual pathologies rather than as social problems, and young people are now blamed for these social issues and treated as criminals rather than as victims—let alone valued as investments in the future. Children have become the enemy within.

A public rhetoric of fear, control, and surveillance presents children as alien, removed from the social contract, and divorced from institutions capable of protecting their rights. Daytime talk show hosts such as Sally Jesse Raphael and Jerry Springer offer endless images of kids out of control—narcissistic, selfish, violent, immoral, sex- and drug-addicted—and public policy reinforces these images by suggesting that the only way to deal with kids is to severely discipline them, even if this means incarcerating them at record levels and in some cases putting them to death. For a third of all minority youth, the future holds the disturbing possibility of either "prison, probation, or some form of supervision within the criminal justice system."[35] Until the recent recession, states were spending more on prison construction than on higher education. Paul Street claims that in Illinois, for every "African-American enrolled in [its] universities, two and a-half Blacks are in prison or on parole. . . . [While] in New York . . . more Blacks entered prison just for drug offenses than graduated from the state's massive university system with undergraduate, masters, and doctoral degrees combined in the 1990s."[36] Under such circumstances, repressive practices cannot be simply justified by the war on terrorism. On the contrary, repressive policies and practices are reinforced and extended through an appeal to national security, but the roots of such undemocratic actions lie in the spreading of neoliberalism and its transformation of the democratic state into the corporate state, and political power largely into a force for domestic militarization and repression. Understanding the relationship between repression

and the war on young people can help critics and activists concep-
tualize the current attack on civil liberties as part of a broader crisis
over the political and ethical importance of the social sphere and
the possibility of upholding the very idea of a democratic future
both nationally and internationally.

   With few exceptions, debates about globalization also seem to
take place in a world without young people. And, yet, the massive
changes prompted by globalization have had a profound affect on
many of the world's 2.9 billion children. In a new world order
marked by deregulation, acceleration, free-flowing global finance,
trade, and capital, short-term gains replace long-term visions. The
search for markets and profits is now buttressed by highly destruc-
tive and sophisticated military technologies that work in conjunc-
tion with new global information systems that overcome the
burden of geographical distance while creating ruling elites that no
longer feel committed or obligated to traditional notions of place,
whether they be towns, cities, states, or nations. Reality TV, with
its social Darwinian logic, supplies the fodder for high television
ratings as it provides global audiences with models of social justice
repackaged as laws of nature, and citizenship as an utterly privatized affair. Neoliberal globalization widens the gap between the
public and the private, on the one hand, and politics and economic
power on the other. Globalization now signals the retreat of
nation-states that once played a significant role in ameliorating the
most brutal features of capitalism. As the nation-state abdicates its
traditional hold on power,[37] it is being replaced by the national
security state, engaged in both fighting the alleged threats from
domestic terrorism—signaled by over-the-top racial profiling and
anti-youth repression—and external terrorism justifying the most
blatant forms of racism and xenophobia directed at Arab and
Muslim populations abroad.

   The consequences of neoliberal globalization can be seen not
only in growing inequalities worldwide in income, wealth, basic
services, and health care, but also in substantial increases in the
exploitation and suffering of millions of young people around the
globe. The fallout is easy to document. As globalization and militarization mutually reinforce each other as an economic policy and
a means to settle conflicts, wars are no longer fought between

soldiers but are now visited upon civilians, and appear to have the most detrimental effects on children. Within the last decade, 2 million children have died in military conflicts. Another 4 million have been disabled, 12 million have been left homeless, and millions more have been orphaned.[38] Increasingly, children are being recruited, abducted, or forced into military service as lighter weapons enable children as young as 12 to be trained as effective killers. But the fruits of modern warfare not only enable children to kill; they also result in their deaths, especially through the proliferation of land mines, which are estimated to kill 8,000 to 10,000 children each year. The International Committee on the Red Cross estimates that "some 110 million land mines threaten children in more than 70 countries" and that they are chillingly effective: "82.5 percent of amputations performed in ICRC hospitals are for land mine victims."[39]

As the leading supplier of arms in the world, the United States bears an enormous responsibility for fueling military conflicts throughout the globe. As reported by the Congressional Research Service, a division of the Library of Congress, American manufacturers in 2000 signed contracts for just under $18.6 billion in weapon sales, going primarily to developing countries. Empire in this instance is not simply about the power pretensions of an imperial presidency, it is also about neoliberal policies that feed corporate profits through the selling of arms that cripple and kill children. Hence, it should come as no surprise to learn that in the age of empire, domestic markets even in the United States are no longer at a safe remove from the scorched earth policies and consequences of arms manufacturers. For instance, the United States ranks worst among industrialized nations in the number of children killed by guns, with over 50,000 American youth killed since 1979. Globalization is not only about the emergence of new technologies, the consolidation of corporate power, and the flow of financial capital, it is also about the intersection of profits and militarization—and the killing and maiming of poor children at home and abroad. Under such circumstances, it makes more sense for leftist critics to address the issue of globalization and Bush's "Axis of Evil" moralism by exposing the administration's hypocrisy in promoting the conditions for military conflict all over the globe.

In spite of what the neoliberal cheerleaders claim, globalization is not simply about creating "free trade" and opening markets. In actuality, it refers to "advancing . . . corporate and commercial interests"[40] through the internationalization of armed conflict and globalization policies fueled by the incessant need for profits—whatever the human costs.

The division of labor and exploitation promoted through neoliberal globalization has given new meaning to Manuel Castells's pronouncement that the primary labor issue in the new information age "is not the end of work but the condition of work."[41] The search for cheap labor, the powerlessness of children, and the 120 million children who are born poor each year create fertile conditions for multinational corporations to profit by hiring children, largely in developing countries. The International Labor Office estimates that 120 million children between the ages of 5 and 14 are compelled to work full-time, often under harsh and inhumane conditions, and that if part-time work is included the figure reaches 250 million.[42] Children are engaged in a variety of forms of labor ranging from domestic servants and shoe production to brickmaking and agricultural work. Many children are either injured or killed on the job, with the number of annual injuries estimated at 70,000. Not all children are exploited by being paid substandard wages for their work; the most unfortunate, and generally the most destitute, are forced into bonded slavery in order to pay off their family loans, or are sold outright on the market by their families in the hopes of bringing in additional income. Many of these children are forced into prostitution, domestic service, or put on the streets as beggars. Prostitution, in particular, has become a substantial growth market fueled by the globalization of child pornography rings, which are largely organized and promoted through the Internet and other global circuits of power such as organized sex tours. While the figures on this illicit trade are difficult to establish, it has been estimated by the Center for Protection of Children's Rights that more than a million children enter into prostitution each year, many with or contracting HIV. Child prostitution is also on the rise in the United States, with an estimated 100,000 to 300,000 children exploited through prostitution and pornography.[43] In spite of the crisis that children are facing all over the world with respect to the violation of their rights and their bodily dignity, the United States has both refused to address critically the myriad ways in which it contributes to turning innocent children into victims through policies that strip countries of their public services, resources, and revenue, and declined to ratify a number of international treaties designed specifically to improve the quality of life for the world's children. The message that is unabashedly put forth by the Bush administration about children both at home and abroad is that it has little regard for the bodies and minds of young people. Market-driven politics suggests that young people under neoliberalism do not merit human rights, social justice, health care, environmental protection, or minimal social provisions. How else to explain the refusal of the Bush administration to sign or ratify the United Nations Convention on the Rights of the Child (passed in 1989), the small arms treaty, and the land mines treaty? The Bush administration's disregard for the health and welfare of children is also evident in its opposition to the Kyoto Protocol on climate change, the International Plan for Cleaner Energy, the protocol for the Biological Weapons Convention, and the International Covenant on Economic, Social and Cultural Rights.[44]

Intellectual Responsibility and Civic Courage

If we are to think our way to a future different from the insensate scenario of unlimited warfare that has been prescribed for us, then culture needs to imagine alternative forms that are not even dreams at present—produced for a public that extends beyond the initiates, and "political" in the sense of relevant to worldly affairs—with confidence that a truly unforced cultural project will be free of both the fundamentalist intolerance and the commercial libertinism that, from partial perspectives, are now so feared.[45]

As a public sphere that prepares youth for the future as well as shapes it, higher education is deeply implicated in how it relates to broader social, political, and economic forces that bear down on youth. As the subject and object of learning, youth provide faculty and administrators with a political and moral referent for addressing the relationship between knowledge and power, learning and social change, and values and classroom social relations as they bridge the gap between the diverse public spheres that youth inhabit and the university as a site of socialization and political engagement. The overwhelming presence of middle- and upper-class youth in the university raises important questions about the role of higher education in furthering and reproducing those divisions of labor between the rich and the poor that are made visible not only in the class and racial imbalances of most student populations, but also in a range of social relations outside the university. Educators would do well in their own teaching to address how higher education furthers class, racial, and gender divisions that resonate with dominant modes of exclusion and discrimination, and that are accentuated under neoliberalism. Surely, if educators have a responsibility to fight against those forces that undermine the university's claims to providing a quality education for all students, they would have to address the increasing corporatization of university life and its effects on the university as a democratic public sphere. For instance, they might want to address the role of neoliberalism in raising tuition rates, deepening the fiscal crisis of the state, contributing to massive cuts in funding for higher education, and the growing exploitation of adjuncts and graduate students in many universities and colleges.

One of the challenges that academics face as engaged intellectuals centers around recovering the language of sociality, agency, solidarity, democracy, and public life as the basis for creating new conceptions of pedagogy, learning, and governance. Part of this effort demands creating new vocabularies, experiences, and roles that allow students to develop a sense of leadership, to question what it is they have become within existing institutional and social formations, and "to give some thought to their experiences so that they can transform their relations of subordination and oppression."[46] Samir Amin rightly argues that it is the absence of social values such as generosity and human solidarity that "reinforce[s] submission to the dominating power of capitalist ideology."[47] And yet, it is precisely through a focus on the obligations of adult society to young people that such values become concrete and persuasive. It is often difficult for adults and students to dismiss the suffering of young people as a matter of individual character or to reduce their plight to the realm of the family or private sphere—the depoliticizing strategy of choice used by social conservatives and neoliberals. The plight of children provides a powerful stimulus for public consciousness. Young people offer a compelling referent for a pedagogy of disruption, social criticism, and collective change because their suffering and hardships offer the pedagogical promise of both a public hearing and an opportunity to connect a range of issues and problems that are often addressed in isolation, a subtle way of identifying grievances without inquiring into their social and political roots. More than any other group, they provide a compelling reason for challenging the moralisms and policies of conservatives while simultaneously opening up the possibility to create new ethical discourses, modes of agency, and forms of advocacy. Young people are one of the few causes left for reclaiming a future that does not imitate the present, a future that makes good on the promise of new models of human association and pedagogy based on democratic values, and a radical transformation of the existing inegalitarian structures of political power and economic wealth. A social analysis of the crisis of youth is not only important for its own sake, but also because it points to much broader analysis in that the various forms of oppression that young people experience directly undermine the dominant and traditional justifications for class, racial, sexual, and gendered divisions in society. For instance, the long neglected discussion of class becomes more visible and poignant when analyzing the inhumane and class-specific effects of George W. Bush's economic stimulus policies, which offer huge tax cuts for the rich while driving the United States into massive deficits and debts that will cut many viable public services and social programs for many children in addition to saddling the next generation "with nothing but a mountain of debt."[48] While it is crucial for educators and others to make clear that Bush's budget policies will do little to help the poor, elderly, homeless, and disabled, such criticism becomes more powerful when children are included as a crucially affected population. It is important to shed light on the fact that the effects of the tax cuts for many children will be devastating—with over 50,000 kids eliminated from after school programs, 8,000 homeless young people cut from vital education programs, and over 33,000 children dropped from child care. It is both important to highlight the effects these cuts are having on children and politically necessary to use such cuts as a ruthless example of class warfare in the broader sense. For example, public discussions about Bush's handling of the economy is often shrouded in a statistical language that masks the social damage of an economic policy that has resulted in over 5.6 million young people unemployed as of 2003—with the youth unemployment rate in some cities such as Wichita, Kansas, reaching as high as 50 percent. The same can be said about Bush's occupation of Iraq. Bush talks about building schools in Iraq, but does nothing to prevent public school systems in the United States from shortening the school year, laying off many needed teachers, and dropping programs in music and art. The government has also failed to provide much-needed financial resources to rebuild the decaying physical infrastructure of a school system largely built in the 1950s. How might an analysis of the state of today's youth be used to raise questions about the ethical, political, and economic priorities of a country that spends more on beauty products than on education?[49] What does it say about a political system that neither calls into question such shameful priorities nor does anything to challenge them?

Young people provide a crucial lens through which hegemony can be analyzed, compassion mobilized, and politics engaged beyond local interests and national boundaries. Ideological domination in this instance does not simply refer to the ideas, discourses, or images that represent young people in particular ways, but also the way in which they actually experience the different modalities of power and powerlessness as an empirical reality within particular class and racial formations marked by deep inequalities of power within and across national boundaries.[50] Young people are born into the existing social order and cannot be blamed so easily for the conditions of poverty, racism, and daily violence that produce inadequate health care, education, and housing for the most defenseless and least powerful. The oppression of young people is crucial for public intellectuals to address because it is the fundamental lie at the heart of neoliberalism and its falsely "utopian" notion of the future.

Though we take up this issue in greater detail in chapter 6, we want to stress that children should be the focus of renewed critical discussion about the long-term consequences of current policies and social practices because they provide a powerful referent for decoding and understanding the suffering of others. They evoke compassion and cause moral unease, making it possible to reassert the importance of the social sphere, civic engagement, political imagination, and a culture of questioning.[51] Focusing on the social position of children opens up an ethical and political space for educators to translate alleged individual problems into public considerations and public considerations into personal concerns, particularly as progressives grapple with questions of politics, power, social justice, and public consciousness. The plight of young people must play a central role in rearticulating the promise of critical citizenship and reaffirmation of a social contract that embraces democratic values, practices, and identities while challenging the limitations of those devisive relations and alienated identities produced by neoliberalism. Making visible the suffering and oppression of young people cannot help but challenge the core ideology of neoliberalism.

Educators need a new language in which young people are central to a transformative notion of pedagogy conceived in terms of social and public responsibility. The growing attack on youth in American society may say less about the reputed apathy of the populace than about the bankruptcy of old political languages and orthodoxies, pointing up the need for new vocabularies and visions for clarifying our intellectual, ethical, and political projects, especially as they work to inject agency, ethics, and meaning back into politics and public life. In the absence of such a language, as well as the social formations and public spheres that make democracy and justice operative, politics becomes narcissistic and caters to the widespread mood of pessimism and the cathartic allure of the spectacle. In addition, public service and government intervention are sneered at as either bureaucratic or a constraint upon individual freedom. To give new life to a substantive democratic politics, educators must address the issues of how people learn to be political agents and what kind of educational work is necessary within what kind of public spaces to enable them to use their full intellectual resources to critique existing institutions and struggle to make freedom and autonomy a reality for as many people as possible. As critical educators, we are required to understand more fully why the tools we used in the past feel awkward in the present, often failing to respond to problems now facing the United States and other parts of the globe. Educators face the challenge posed by the failure or absence of oppositional discourses and the disorganization of dissent to bridge the gap between how society represents itself and how and why individuals fail to understand and critically engage such representations in order to intervene in the oppressive social relationships that representations often legitimate. Educators need a language adequate to the demands of a global public sphere. They need to understand how the local affects the global public sphere and vice versa. Most important, they need a language that rises above a politics of castigation. Such a language must reclaim the voices and experiences of those critical traditions and social movements that negotiated "between the real and the ideal in protest against the societies and power structures in which they emerge."[52] Educators need a new language for expressing global solidarity as well as an understanding of the political and pedagogical strategies necessary to create a global public sphere where such solidarities become possible.

At a time when civil liberties are being destroyed, massive tax cuts are being given to the rich, and public services gutted, the nation squanders its resources in maintaining military control of Iraq. At the same time, public institutions and goods all over the globe are under assault by the forces of a rapacious global capitalism; there is a sense of urgency that demands that academics develop new modes of resistance and collective struggle buttressed by rigorous intellectual work, social responsibility, and political courage.

As theorists such as Pierre Bourdieu, Noam Chomsky, Howard Zinn, Arundhati Roy, and Edward Said have reminded us, intellectuals have a special responsibility to use their talents to address crucial social issues, present alternative narratives that make dominant power accountable, and offer alternative strategies of intervention to realize a democratic future. In part, this means not only offering a critical analysis of representations that serve dominant power and legitimate the status quo, but also making visible those issues that exist outside of dominant discourses and the social conditions that produce them. As intellectuals, academics need to make public the experiences of those whose voices are either excluded from public debate, or, when they are heard, rarely carry any sort of authority.

Academics need to connect their work to a larger public and assume a measure of responsibility in naming, struggling against, and alleviating human suffering. This means working with others to produce knowledge in a variety of public spheres that can address those forms of social suffering, relations of power, and cultural formations (such as media concentrations) that pose a threat to democracy. As mentioned earlier, academics need to reject the cult of professionalism and assume the role of citizen-scholars, which means, as Edward Said pointed out, maintaining "a kind of coexistence between the necessities of the field and the discipline of the classroom, on the one hand, and of the special interest that one has in it, on the other, with one's own concerns as a human being, as a citizen in a larger society."[53] Such a recognition places a particularly important demand upon academics, who increasingly depoliticize the very possibility of politics as they retreat into arcane discourses and specializations—or exhibit a moral indifference to the outside world. Academics cannot collapse politics into a dehistoricized, text-centered pedagogy that "approaches the social world as if it was a text and reduces the role of the intellectual to a mere reader of texts."[54] One imperative of a critical pedagogy is to offer students opportunities to become aware of their potential and responsibility as individual and social agents to expand, struggle over, and deepen democratic values, institutions, and identities. They must help students unlearn the presupposition that knowledge is unrelated to action, conception to implementation, and learning to social change. Knowledge, in this case, is about more than understanding; it is also about the possibilities of self-determination, individual autonomy, and social agency. Rather than consolidate authority, academics need to make it accountable, tempering a reverence for power and authority with a deep distrust of its motives and effects. Academics need to reclaim not only their intellectual courage but a sense of ethical responsibility.

Connecting academic work to social change should not be summarily dismissed either as partisan or burdensome. Arundhati Roy has argued that there are times in a nation's history when its political climate makes it imperative for intellectuals to take sides. Given the assault being waged by the Bush administration on public services, the welfare state, the environment, workers, civil rights, and democracy itself, we believe that this is a crucial time in American history for academics to make their voices felt in the struggle to reclaim democracy from the market fundamentalists, powerful corporate interests, and evangelical neoconservatives.[55] Pierre Bourdieu insightfully suggests that the time is right for intellectuals to assume responsibility for creating an international social movement that would exercise real influence on transnational corporations, nation-states, and nongovernmental agencies.[56] According to Bourdieu, neoliberalism in its current forms is so ruthless in its destruction of public goods, everyday social protections, meaningful labor, and the environment that it is bringing together academics, workers, students, farmers, consumers, and activists into new alliances.

Intellectuals—academic and non-academic alike—have a special responsibility to enable the conditions for such protests to offer opportunities for new social actors and constructive modes of collective action and political intervention leading to new social policies, rather than allowing protests to degenerate into what Alain Touraine has called the politics of "pointless denunciations."[57] Intellectuals at all costs must fight against the mythic assumption of the neoliberal order that there are no alternatives, and in doing so, resist the slide into cynicism and apathy with a new political language and vision, one marked by a discourse of critique and possibility. Such a discourse must move beyond analyzing only the crushing effects of domination, recognizing "that individual and groups [be regarded] as potential actors and not simply as victims who are either in chains or being manipulated."[58] At stake here is the need not only to combat a debilitating cynicism but also to capture the complexity of relations of power and resistance, recognizing that there are multiple sites where social actors can provide individual criticism and engage in the arduous task of mobilizing social movements. For instance, the antiwar movement,

antisweatshop movement, Living Wage campaigns, and global justice movements have brought together a number of students across
the country who are creating alternative campus media to get their
voices heard, both to reach other students and to affect larger public discourses. Increasingly, such students are reaching out to other
groups, such as trade unionists, and building wider alliances.
Academics need to provide financial and intellectual support for
progressive student publications and forms of social activism—not
only because students represent an important political movement
for social change, but also because they cannot allow conservative
organizations like the Collegiate Network and the Leadership
Institute to pour money into a network of campus newspapers in
order to push a generation of college students toward the right and
away from broader progressive alliances. Academics also need to
address those policies championed by conservatives that include
eliminating affirmative action and using trustees to support right-
wing educational reforms, as well as efforts to convince state legislators to pass laws that would, under the rubric of promoting
ideological diversity, "encourage—if not require—colleges to hire
faculty and invite speakers with conservative views."[59] And these
are only two instances where such resistance can be acknowledged
and supported politically and pedagogically.

The time has come for intellectuals to distinguish caution from
cowardice and recognize that their obligations extend beyond
deconstructing texts or promoting a culture of questioning. These
are important, but they do not go far enough. We also need to link
knowing with action, and learning with social engagement: we
must fulfill the responsibilities that come with teaching students to
fight for an inclusive and radical democracy by recognizing that
education is not just about understanding, however critical, but
also about providing the conditions for addressing the responsibilities we have as citizens to others, especially those future generations
who will inherit our mistakes. It is also crucial for educators to recognize that matters of responsibility, social action, and political intervention do not simply develop out of social critique, but also forms
of self-critique. Hence, they should treat the relationship between
knowledge and power critically, exercising a certain self-reflexivity
about its effects, about what it means to take seriously matters of

individual and social responsibility in addressing those forms of human suffering that are produced by inequalities that undermine any viable democracy. Neoliberalism not only places capital and market relations in a no-man's-land beyond the reach of compassion, ethics, and decency; it also undermines those basic elements of the social contract in which self-reliance, confidence in others, and a trust in the longevity of public institutions provide the basis for individual autonomy, social agency, and critical citizenship. The struggle over the social contract is part of a broader struggle over education, power, and democracy in which young people are seen as the most valuable resource for ensuring an inclusive and just society that will help guarantee them a future of justice, dignity, and security.

If educators are to address the urgency of the crisis that links youth and democracy, they will have to betray those dominant intellectual traditions that divorce academic life from politics, reduce teaching to forms of instrumental rationality that largely serve market interests, and remove the university from those democratic values that hold open the promise of a better and more humane life. They will also have to reject the current "Enronization" of public life, and take the government back from the religious zealots, neoconservative ideologues, and market fundamentalists who are truly trampling on constitutional freedoms, collapsing the rule of democracy into the rule of capital, and making the world a better place for a very small group of powerful individuals and wealthy corporations. These are dangerous times, and it behooves all of us in higher education to wake up and begin to step forward collectively in order to stop this slide into the abyss of a new kind of authoritarianism.

Chapter 3

Cultural Studies and Critical Pedagogy in the Academy

If you accept my definition that this is really what Cultural Studies has been about, of taking the best we can in intellectual work and going with it in this very open way to confront people for whom it is not a way of life, for whom it is not in any probability a job, but for whom it is a matter of their own intellectual interest, their own understanding of the pressures on them, pressures of every kind, from the most personal to the most broadly political—if we are prepared to take that kind of work and to revise the syllabus and discipline as best we can, on this site which allows that kind of interchange, then Cultural Studies has a very remarkable future indeed.

—Raymond Williams[1]

The Promise of Cultural Studies

Cultural studies as a field seems to have passed into the shadows of academic interests. Globalization and political economy have become the privileged concerns of left academics as we move into the new millennium. While we do not want to suggest that the new-found interest in globalization and political economy is unwarranted, we do want to stress that cultural studies's long-standing interest in the interrelationship between power, politics, and culture is much too important at present to be dismissed as the passage of another academic fashion. Matters of agency, consciousness, pedagogy, rhetoric, and persuasion are central to any public discourse about politics, not to mention education itself. In fact, as we argue below, culture is a central sphere of politics; it is the one site that offers both a language of critique and possibility, a sphere in which matters of economy, institutional power relations, globalization, and politics can be recognized, critically understood, and collectively engaged. Hence, the promise of cultural studies, especially as a fundamental aspect of higher education, does not reside in a false opposition between culture and material relations of power, but in a project that bridges these concerns as part of a larger transformative and democratic politics in which matters of pedagogy and agency play a central role.

In the last few decades, a number of critical and cultural studies theorists such as Stuart Hall, Lawrence Grossberg, Douglas Kellner, Meghan Morris, and Richard Johnson have provided valuable contributions to our understanding of how culture deploys power and is shaped and organized within diverse systems of representation, production, consumption, and distribution. Particularly important to such work is an ongoing critical analysis of how symbolic and institutional forms of culture and power are mutually entangled in constructing diverse identities, modes of political agency, and the social world itself. From this perspective, material relations of power and the production of social meaning do not cancel each other out but constitute the precondition for all meaningful practices. Culture is recognized as the social field where goods and social practices are not only produced, distributed, and consumed, but also invested with various meanings and ideologies that have widespread political effects. For example, media and popular culture have an enormous effect in shaping everyday assumptions about the alleged relationship between race and the culture of criminality. Culture is partly defined as a circuit of power, ideologies, and values in which diverse images, texts, and sounds are produced and circulate; identities are constructed, inhabited, and discarded; agency is manifested in both individualized and social forms; institutions produce and constrain social practices; and discourses are created that make culture itself the object of inquiry and critical analyses. Rather than being static, the substance of culture and everyday life—knowledge, goods, social practices, and contexts—repeatedly mutates and is subject to ongoing changes and interpretations.

Following the work of Antonio Gramsci and Raymond Williams, many cultural theorists acknowledge the primacy of culture's role as an educational site where identities are being continually transformed and power is enacted. Learning assumes a political dynamic as it becomes not only the condition for the acquisition of agency, but also the sphere for imagining oppositional social change. As both a space for the production of meaning and social interaction, culture is viewed by many contemporary theorists as an important terrain in which agency, identity, and values are neither unchanging nor always in place but subject to negotiation and struggle and open to new democratic transformations— though always within varying degrees of iniquitous power relations. Rather than a mere reflection of larger economic forces or as simply the "common ground" of everyday life, culture is to many advocates of cultural studies both a realm of contestation and of utopian possibility, a space in which an emancipatory politics can be fashioned that "consists in making seem possible precisely that which, from within the situation, is declared to be impossible."[2]

Culture is where exchange and dialogue become crucial as an affirmation of a democratically configured social space in which the political is actually taken up and lived out through a variety of intimate relations and social formations. Far from being exclusively about matters of representation and texts, culture becomes a field, event, and performance in which identities and modes of agency are configured through the mutually determined forces of thought and action, body and mind, and time and space. Culture is the public space where common matters, shared solidarities, and public engagements provide the fundamental elements of democracy. Culture is also the pedagogical and political ground in which shared solidarities and a global public sphere can be imagined as a condition of democratic possibilities. In this perspective, culture offers what Jonathan Rowe calls a "Temporal commons"—"a pool of time available for the work that the market neglects"[3]—in which to address the radical demand of a pedagogy that allows critical discourse to confront the inequities of power and promote the possibilities of shared dialogue and democratic transformation. And culture's urgency, as Nick Couldry observes, resides in its possibilities for linking politics to matters of individual and social agency as they are lived out in particular democratic spheres, institutions, and communities. He writes:

> For what is urgent now is not defending the full range of cultural production and consumption from elitist judgement but defending the possibility of any shared site for an emergent democratic politics. The contemporary mission of cultural studies, if it has one, lies not with the study of "culture" (already a cliche of management and marketing manuals), but with the fate of a "common culture," and its contemporary deformations.[4]

Cultural studies theorists have greatly expanded our theoretical understanding of the ideological, institutional, and performative workings of culture, but as important as this work might be, it does not go far enough—though there are some exceptions, such as the work of Stanley Aronowitz, Richard Johnson, Doug Kellner, Lawrence Grossberg, bell hooks, and Nick Couldry—in connecting the most critical insights of cultural studies with an understanding of the importance of critical pedagogy, particularly as part of a larger project for expanding the possibilities of a democratic politics, the dynamics of resistance, and the capacity for social agency. For too many theorists, pedagogy often occupies a limited role theoretically and politically in configuring cultural studies as a form of cultural politics.[5] For instance, when invoked as a political practice, pedagogy is either limited to the role that oppositional intellectuals might play within academia or it is reduced almost entirely to forms of learning that take place in schools. Even when pedagogy is related to issues of democracy, citizenship, and the struggle over the shaping of identities and identifications, it is rarely linked to a broader public politics—a larger attempt to explain how learning takes place outside of schools or what it means to assess the political significance of understanding the broader educational force of culture in the new age of media technology, multimedia, and computer-based information and communication networks. For most cultural studies theorists, pedagogy is limited to what goes on in schools, and the role of theorists who take up pedagogical concerns is largely reduced to doing or teaching cultural studies within the classroom.

Within this discourse, cultural studies becomes available as a resource to educators who can then teach students how to look at the media (industry and texts), analyze audience reception, challenge rigid disciplinary boundaries, critically engage popular culture, produce alternative ways of engaging the world, or use cultural studies to reform the curricula and challenge disciplinary formations within public schools and higher education. For instance, Shane Gunster has argued that the main contribution that cultural studies makes to pedagogy "is the insistence that any kind of critical education must be rooted in the culture, experience, and knowledge that students bring to the classroom."[6] While this is an important insight, it has been argued in enormously sophisticated ways for over 50 years by a host of progressive educators that include John Dewey, Maxine Greene, and Paulo Freire. But the problem lies not in Gunster's unfamiliarity with such scholarship, but in his willingness to repeat the presupposition that the classroom is the exclusive site in which pedagogy becomes a relevant object of analysis. If he had crossed the disciplinary boundaries that he decries in his celebration of cultural studies, he would have found that educational theorists such as Roger Simon, David Trend, and others have expanded the meaning of pedagogy as a political and moral practice and extended its application far beyond the classroom, while also attempting to combine the cultural and the pedagogical as part of a broader notion of political education and cultural studies.[7]

Many cultural studies theorists, such as Richard Johnson, Stuart Hall, and Michael Green, have rightly suggested that cultural studies has an important role to play in helping educators rethink, among other things, the nature of pedagogy, knowledge, the purpose of schooling, and how schools are impacted by larger social forces.[8] Gunster takes such advice seriously but fails to understand its limits, and by doing so repeats a now familiar refrain among

critical educational theorists about connecting pedagogy to the histories, lived experiences, and discourses that students bring to the
classroom. In spite of the importance of bringing matters of culture
and power to the schools, we think that too many cultural studies
theorists are remiss in suggesting that pedagogy is primarily about
schools, and by implication that the intersection of cultural studies
and pedagogy has little to do with theorizing the role that pedagogy might play in linking learning to social change outside of traditional spheres of schooling.[9] Pedagogy is not simply about the
social construction of knowledge, values, and experiences; it is also
a practice embodied in the interactions among educators, audiences, texts, and institutional formations. Pedagogy, at its best,
implies that learning takes place across a spectrum of social practices and settings in society. As Roger Simon observes, pedagogy
points to the multiplicity of sites in which education takes place
and offers the possibility for a variety of cultural workers

> to comprehend the full range of multiple, shifting and overlapping
> sites of learning that exist within the organized social relations of
> everyday life. This means being able to grasp, for example, how
> workplaces, families, community and institutional health provision,
> film and television, the arts, groups organized for spiritual expres-
> sion and worship, organized sport, the law and the provision of legal
> services, the prison system, voluntary social service organiza-
> tions, and community based literacy programs all designate sets of
> organized practices within which learning is one central feature and
> outcome.[10]

In what follows, we want to argue that pedagogy is central to any
viable notion of cultural politics and that cultural studies is key to
a critical notion of pedagogy, especially as a practice and object of
scholarship in higher education. Moreover, it is precisely the intersection of diverse traditions in cultural studies and pedagogy that
presents the possibility for making the pedagogical more political
for cultural studies theorists and the political more pedagogical for
educators. We are particularly concerned about how the intersection of cultural studies, pedagogy, and politics can be read as an
effort to redefine the role of academics as public intellectuals,

higher education as a crucial public sphere for educating students to address matters vital to a democratic society, and pedagogy as enabling both a culture of questioning as well as strategic interventions into those practices, structures, and struggles that connect learning to public life.

Rethinking the Importance of Cultural Studies for Educators

Our own interest in cultural studies emerges out of an ongoing project to theorize the regulatory and emancipatory relationship among culture, power, and politics as expressed through the dynamics of what we call public pedagogy. Such a project concerns, in part, the diverse ways in which culture functions as a contested sphere over the production, distribution, and regulation of power and how it operates both symbolically and institutionally as an educational, political, and ideological force. Drawing upon a long tradition in cultural studies, we take up culture as constitutive and political—not only reflecting larger forces but also constructing them. Culture not only mediates history, it shapes it. We argue that culture is the primary terrain for realizing the political as an act of social intervention, a space in which politics is pluralized, recognized as contingent, and open to many formations.[11] It is a crucial terrain for rendering visible both the global circuits that now control material relations of power and deploy representations and meanings through which politics is expressed, lived, and experienced. Culture in this view is the ground of not only power and politics but also a crucial domain of contestation and accommodation—increasingly characterized by the rise of megacorporations and new technologies, which are transforming radically the traditional spheres of the economy, industry, society, and everyday life. We are referring not only to the development of new information technologies, but also the enormous concentration of ownership and power among a limited number of corporations that now control a diverse number of media technologies and markets. Culture now plays a central role in producing narratives, metaphors, and images that exercise a powerful educational force over how people think of themselves and their relationship to others. From our perspective, culture is the primary sphere in which individuals, groups, and institutions translate the diverse and multiple relations that mediate between private life and public concerns. It is also the sphere in which the possibilities for dissent, dialogue, and translation are under assault, particularly as the forces of neoliberalism dissolve public issues into utterly privatized and individualistic concerns.

Central to our work in cultural studies is the assumption that the primacy of culture and power be organized through an understanding of how the political becomes pedagogical, and particularly how private issues are connected to larger social conditions and collective forces; that is, how the very processes of learning constitute political mechanisms through which affective investments, subject positions, and everyday relations are given form and meaning within and through collective conditions and those larger forces that constitute the realm of the social. In this context, pedagogy is no longer restricted to what goes on in schools, but becomes a defining principle of a wide range of cultural apparatuses engaged in what Raymond Williams has called "permanent education," which we discussed in chapter 1. Williams rightfully believed that education in the broadest sense—as something that is not limited to schools—plays a central role in any viable form of cultural politics. Williams argued that any viable notion of critical politics would have to pay closer "attention to the complex ways in which individuals are formed by the institutions to which they belong, and in which, by reaction, the institutions took on the colour of individuals thus formed."[12] Williams also foregrounded the crucial political question of how agency unfolds within a variety of cultural spaces structured by unequal relations of power.[13] He was particularly concerned about the connections between pedagogy and political agency, especially in light of the emergence of a range of new technologies that radically increased the amount of information available to people and the speed at which they could access it, while at the same time constricting the substance and ways in which such meanings entered the public domain. The realm of culture for Williams took on a new role in the latter part of the twentieth century because economic power and its networks of control now exercised more influence than ever before in shaping how identities are produced, desires mobilized, and everyday social relations were shaped.[14] Williams clearly understood that making the political more pedagogical meant recognizing that where and how the psyche locates itself in public discourse, visions, and passions provides the groundwork for agents to enunciate, act, and reflect on themselves and their relations to others and the society.

Following Williams, we want to reaffirm the importance of pedagogy in cultural politics. We also want to comment on some concepts central to cultural studies that are useful not only for thinking about the interface between cultural studies and critical pedagogy, but also for deepening and expanding the theoretical and political horizons of critical pedagogical work in higher education. We believe that pedagogy represents both a mode of cultural production and a type of cultural criticism that is essential for questioning the conditions under which knowledge is produced, values affirmed, affective investments mobilized, and subject positions put into place, negotiated, taken up, or refused. Pedagogy is a referent for understanding the conditions for critical learning and the often hidden dynamics of social and cultural reproduction. Most important, it is the precondition for critical citizenship, social responsibility, and a vibrant and inclusive democracy. As a critical practice, pedagogy's role lies in not only changing how people think about themselves, their relationship to others and the world, but also in energizing students and others to engage in those struggles that further possibilities for living in a more just and fairer society. But like any other body of knowledge that is constantly struggled over, pedagogy must constantly enter into dialogue with other fields, theoretical domains, and emerging scholarly discourses. As diverse as cultural studies is as a field, there are a number of insights it provides that are crucial to educators who use critical pedagogy both in and outside their classrooms.

First, in the face of contemporary forms of political and epistemological relativism, a more politicized version of cultural studies makes a claim for the use of highly disciplined, rigorous theoretical work. Not only does such a position reject the notion that intellectual authority can only be grounded in particular forms of social identity, it also refuses an increasing anti-intellectualism that posits theory as too academic and complex to be of any use in addressing important social and political issues. While many cultural studies advocates refuse to either separate cultural studies from politics or reject theory as too complex and abstract, they also reject the potential insularity of theory as a sterile form of theoreticism, an academicized jargon that is as self-consciously pedantic as it is politically irrelevant. Language, experience, power, ideology, and representation cannot avoid theory, but that is no excuse for elevating theory to an ethereal realm that has no referent outside of its own obtuseness or rhetorical cleverness. Instead, theory can be a resource for connecting cultural studies to those areas of contestation in which it becomes possible to open up rhetorical and pedagogical spaces where we can challenge the actual conditions of dominant power and create the promise of a future that contains a range of democratic alternatives.[15] Theory in this sense does not merely refer to itself but is valued for its ability to open up new horizons of political possibility and to offer strategic interventions for shaping everyday life. As a resource, theory both highlights and interprets, and thereby connects a broad range of institutions and discourses to social practices and the larger society. Lawrence Grossberg clearly articulates this position in his comment on the role of theory within cultural studies. He writes:

> Theory in cultural studies is measured by its relation to, its enablement of, strategic interventions into the specific practices, structures, and struggles characterizing its place in the contemporary world. Cultural studies is propelled by its desire to construct possibilities, both immediate and imaginary, out of its historical circumstances. It has no pretensions to totality or universality; it seeks only to give us a better understanding of where we are so that we can get somewhere else (some place, we hope, that is better—based on more just principles of equality and the distribution of wealth and power), so that we can have a little more control over the history that we are already making . . . A theory's ability to "cut into the real," to use Benjamin's metaphor, is measured by the political positions and trajectories theory enables in response to the concrete contexts of power it confronts. Just like people in everyday life, cultural studies begins to grapple with

and analyze difficult experiences at hand; it draws upon and extends theories to enable it to break into experience in new ways.[16]

Underlying such a public project is a firm commitment to intellectual rigor, social justice, and civic courage. At stake here is a deep regard for matters of compassion and social responsibility aimed at broadening the possibilities for critical agency, racial justice, economic democracy, and the just distribution of political power. Hence, cultural studies theorists often reject the anti-intellectualism, specialization, formalism, and compartmentalism often found in other disciplines. Similarly, such theorists reject both the universalizing dogmatism present in some strands of radical theory as well as a postmodern epistemology that enshrines difference, identity, and plurality at the expense of developing more inclusive notions of the social that bring together historically and politically different forms of struggle. The more progressive strains of cultural studies do not define or value theory and knowledge strictly within sectarian ideological or pedagogical interests. On the contrary, these approaches to cultural studies define theorizing as part of a more generalized notion of freedom, which combines democratic principles, values, and practices with the rights and discourses that build on the histories and struggles of those often marginalized because of race, class, gender, disability, and age. For instance, cultural studies theorist Imre Szeman has looked at the ways in which globalization opens up not only a new space for pedagogy but "constitutes a problem of and for pedagogy."[17] Szeman analyzed the various forms of public pedagogy at work in the rhetoric of newspapers, TV news shows, financial service companies, advertising industries, and the mass media, and how such rhetoric fashions a triumphalist view of globalization. He then offers an analysis of how alternative pedagogies are produced within various globalization protest movements that have taken place in cities such as Seattle, Toronto, and Genoa—movements that have attempted to open up new modes of learning while creating new forms of collective resistance. What is particularly important about Szeman's analysis is how new pedagogical practices of resistance are being fashioned through the use of new media such as the Internet, computers, CD-ROMs, and digital video to challenge official pedagogies and dominant views of globalization.

Second, cultural studies is radically contextual in that the very questions that it asks change in every context. Theory and criticism do not become an end in themselves but are always a response to problems raised in particular contexts, social relations, and institutional formations. How we respond as educators and critics to the spheres in which we work is conditioned by the interrelationship between our own theoretical resources and the worldly, space of publicness that produces distinct problems and conditions particular responses to them. While theory is not a substitute for politics, it does provide the very precondition for a critically self-conscious notion of individual and social agency as the basis for shaping the larger society. Politics as an intervention into public life is in this instance part of a broader attempt to provide a better understanding of how power works in historical and institutional contexts and relations of domination and subordination, while simultaneously opening up the possibility of changing them. Lawrence Grossberg puts it well in arguing that cultural studies must be grounded in an act of doing, which in this case means "intervening into contexts and power. . . . in order to enable people to act more strategically in ways that may change their context for the better."[18] Pedagogy is not an *a priori* set of methods that simply needs to be uncovered and then applied, regardless of the context in which one teaches; instead, it is the outcome of numerous struggles between different groups over how contexts are made and remade, often within unequal relations of power. While educators need to be attentive to the particular context in which they work, they cannot separate such contexts from larger configurations of power, culture, ideology, politics, and domination. As Douglas Kellner and Meenakshi Gigi Durham observe, "pedagogy does not elide or occlude issues of power. . . . Thus, while the distinctive situation and interests of the teachers, students, or critics help decide what precise artifacts are engaged, what methods will be employed, and what pedagogy will be deployed, the socio-cultural environment in which cultural production, reception, and education occurs must be scrutinized as well."[19]

The notion that pedagogy is always contextual does more than simply alert educators to diverse forces at work in shaping any learning context; it also points to the importance of connecting the knowledge that is taught to the experiences that students bring to the classroom. Teachers should be educated about the viability of developing context-dependent learning that takes account of student experiences and their relationships to popular culture and its terrain of pleasure, including those cultural industries that are often dismissed as mere entertainment—a pedagogy deeply at odds with the standardization of knowledge and methods now dominant in educational reform movements. Despite the growing cultural diversity of students in higher education, there are few examples of curricular sensitivity to the multiplicity of economic, social, and cultural factors bearing on students' lives. Even where there has been a proliferation of programs such as ethnic and black studies in higher education since the 1970s (though many are being slowly starved), these are often marginalized in small programs far removed from the high status associated with courses organized around business, computer science, and biotechnology. Cultural studies at least provides the theoretical tools for educators to recognize the important, though not unproblematic, cultural resources of students, and the willingness to affirm and engage them critically as forms of knowledge crucial to the students' sense of identity, place, and history. Equally important, the knowledge produced by students offers educators opportunities to learn from young people and to incorporate such knowledge as an integral part of their own teaching. Yet, there is an important caveat that cannot be stated too strongly.

We are not endorsing a romantic celebration of the relevant knowledge and experience that students bring to the classroom. Nor are we arguing that the larger contexts that frame both the culture and political economy of the schools and the experiences of students be ignored. We are also not suggesting that teaching be limited to the resources that students already have, as much as we are arguing that educators need to find ways to make knowledge meaningful in order to make it critical and transformative. Moreover, by locating students within their various histories, experiences, and values, pedagogy can both raise questions about the strengths and limitations of what students know and grapple with the issue of what pedagogical conditions must be engaged to expand the capacities and skills needed by students to become global citizens and responsible social agents. This is not a matter of making a narrow notion of relevance the determining factor in the curriculum. But it is an issue of connecting knowledge produced in the academy to that which is produced in everyday life; connecting meaning to the act of persuasion; relating schools and universities to broader public sphere; and tying rigorous theoretical work to affective investments and pleasures that students use in mediating their relationship to others and the larger world.

Third, the cultural studies emphasis on transdisciplinary work provides a rationale for challenging how knowledge has been historically produced, hierarchically ordered, and used within disciplines to sanction particular forms of authority and exclusion. By challenging the established academic division of labor, a transdisciplinary approach raises important questions about the politics of representation and its deeply entrenched entanglement with specialization, professionalism, and dominant power relations. Transdisciplinary work often operates at the frontiers of knowledge, and prompts teachers and students to raise new questions and develop models of analysis outside the officially sanctioned boundaries of knowledge and the established disciplines that control them. It also serves a dual function: on the one hand, it firmly posits the arbitrary conditions under which knowledge is produced and encoded, stressing its historically and socially constructed nature and its connection to power and ideological interests. On the other hand, it endorses the relational nature of knowledge, countering the idea that knowledge, events, and issues are either fixed or should be studied in isolation. Transdisciplinary approaches stress both historical relations and broader social formations, while remaining attentive to new linkages, meanings, and possibilities. Strategically and pedagogically, these modes of analysis suggest that though educators may be forced to work within academic disciplines, they can develop transdisciplinary tools to challenge the limits of established fields and contest the broader economic, political, and cultural conditions that reproduce unequal relations of power and inequities and the divisions of labor within academic work. This is a crucial turn theoretically and politically, because transdisciplinary approaches foreground the necessity of bridging the work educators do within the academy to other academic fields—as well as to public spheres outside of the academy. Transdisciplinary work provides opportunities for new alliances within and outside the university. It also provides the tools to work in spaces that cover a range of practices and institutions, including radio, film, news media, entertainment, sports, popular culture, churches, synagogues, and elite cultural spheres. As public intellectuals, educators engage in ongoing public conversations that cut across particular disciplines and reach out to more than one type of audience. Such circumstances require that educators address the task of learning multiple forms of knowledge and skills that enable them to speak critically and broadly on a number of issues to a wide range of publics.

Fourth, in a somewhat related way, the emphasis on the part of many cultural studies theorists to study the full range of cultural practices opens the possibility for understanding a wide variety of new cultural forms that function as the primary educational forces in advanced industrial societies. New electronic technologies and the emergence of visual culture as a primary educational force offer new opportunities for teachers and students to engage ways of knowing that simply do not correspond to the long-standing traditions and officially sanctioned rules of disciplinary knowledge or the onesided academic emphasis on print culture. The scope and power of visual culture, for instance, warrant that educators become more reflective about engaging both the production, reception, and situated use of new technologies, popular texts, and diverse forms of visual media and how they structure social relations, values, particular notions of community, the future, and varied definitions of the self and others. Texts in this sense do not merely refer to the culture of print, but to all those audio, visual, and electronically mediated forms of knowledge that have prompted a radical shift in the ways in which knowledge is produced, received, and consumed. Recently, some of our work has focused on the ways in which Disney's corporate culture—its animated films, radio programs, theme parks, and Hollywood blockbusters—functions as both a nontraditional site of pedagogy and an expansive teaching machine, which appropriates media and popular culture in order to rewrite public memory and offer young people an increasingly privatized and commercialized notion of citizenship.[20] We have also addressed how Hollywood films constitute a new form of pedagogical address and exercise a powerful educational force both within and outside of the United States.[21]

Contemporary youth do not simply rely on the culture of the book to construct and affirm their identities; instead, they are faced with the daunting task of negotiating a decentered, media-based cultural landscape no longer caught in the grip of either print technology or closed narrative structures.[22] We believe that educators and other cultural workers cannot critically understand and engage the shifting attitudes, representations, and desires of new generations of youth strictly within the dominant disciplinary configurations of knowledge and practice and traditional forms of pedagogy. Educators require a more expansive view of knowledge and pedagogy that provides the conditions for students to engage popular, media, and mass culture as serious objects of social analysis, learning how to read them critically through specific strategies of understanding, engagement, and transformation. This notion embodies a view of literacy that is multiple and shifting rather than singular and fixed. The traditional emphasis on literacy must be reconfigured in order for students to learn multiple literacies rooted in a mastery of diverse symbolic domains. Similarly, it is not enough to educate students to be critical readers in these various areas; they must also become cultural producers—especially if they are going to create new, alternative public spheres in which official knowledge and its one-dimensional configurations can be challenged. Students must learn how to use the new electronic technologies as well as how to think about the dynamics of cultural power and how it works on and through them, so that they can build alternative cultural spheres in which such power is shared and used to promote noncommodified values rather than simply mimic corporate culture and its underlying transactions. Many cultural studies theorists are well aware of the importance of addressing the social forms in which young people gain a sense of identity and understanding of the world. But the pedagogical task here must go further to include addressing youth-oriented organizations, speaking a language and presenting a vision that resonates with the concerns of youth, all of which are inseparable from the broader issues of democracy and social justice.[23]

Fifth, cultural studies provocatively stresses analyzing public memory not as a totalizing narrative, but as a series of ruptures and displacements. Historical learning in this sense is not about constructing a linear narrative but about blasting history open, rupturing its silences, highlighting its detours, acknowledging the manner of its transmission, and recapturing its concern with human suffering, struggles, values, and the legacy of the often unrepresentable or misrepresented. History is not an artifact to be merely transmitted, but an ongoing dialogue and struggle over the relationship between representation and agency, material relations of power and maps of meaning. The emphasis on struggle makes clear the discursive nature of historical narrative, while also situating history within a notion of contingency that removes it from offering any certainties. This is not meant to imply that some histories are not more truthful and accurate than others. The claim to historical accuracy warrants important concerns about matters of argument, evidence, logic, and methodology. What we are suggesting is that history is constructed through narratives that cannot free themselves from their own social conditioning, and that while some historical accounts may offer important lessons, they offer no guarantees. James Clifford argues that history should "force a sense of location on those who engage with it."[24] This means challenging official narratives of conservative educators such as William Bennett, Lynne Cheney, Diane Ravitch, and Chester Finn, for whom history is both eclectic and standardized, primarily about recovering and legitimating selective facts, dates, and events. In contrast, a pedagogy of public memory is about making connections that are often hidden, forgotten, or willfully ignored. It is also about being aware of how history is shaped for us in museums, schools, the media, and a host of other areas. History and public memory can never be allowed to congeal "into a singular, salvational meaning."[25] Public memory in this sense becomes not an object of reverence but an ongoing subject of debate, dialogue, and critical engagement. Public memory is also about critically examining one's own historical location amid relations of power, privilege, or subordination. Engaging history, as has been done repeatedly by progressive intellectuals such as John Hope Franklin, Howard Zinn, and Noam Chomsky, means analyzing how knowledge is constructed through its absences. Public memory as a pedagogical practice functions, in part, as a form of critique that addresses the fundamental inadequacy of official knowledge in representing marginalized and oppressed groups; it also, as John Beverly points out, reveals the deep-seated injustices perpetrated by institutions that perpetuate such knowledge; and it affirms the need to transform such institutions in the "direction of a more radically democratic nonhierarchical social order."[26]

Sixth, cultural studies theorists are increasingly paying attention to their own institutional practices and pedagogies.[27] They have come to recognize that pedagogy is deeply implicated in how power and authority are employed in the construction and organization of knowledge, desires, values, and identities. Such a recognition has produced a new self-consciousness about how particular forms of teacher authority, classroom knowledge, and social practices are used to legitimate particular values and interests within unequal relations of power. Questions concerning how pedagogy works to construct knowledge, meaning, desire, and values not only provides the conditions for a pedagogical self-consciousness among teachers and students, but also foregrounds the recognition that pedagogy is a moral and political practice that cannot be reduced to an *a priori* set of skills or techniques. Rather, pedagogy is defined as a social practice that must be accountable ethically and politically for the stories it produces, the claims it makes on public memories, and the images of the future it deems legitimate. As both an object of critique and a method of cultural production, critical pedagogical practices cannot hide behind claims of objectivity, and should work, in part, to link theory and practice in the service of organizing, struggling over, and deepening democratic freedoms. In the broadest sense, critical pedagogy should offer students and others—outside of officially sanctioned scripts—the historically and contextually specific knowledge, skills, and tools they need to both participate in, govern, and change when necessary those political and economic structures of power that shape their everyday lives. Needless to say, such tools are not simply given, but are the outcome of debate, dialogue, and engagement across a variety of public spheres.

While this list is both schematic and incomplete, it points to a number of important theoretical considerations that can be appropriated from the field of cultural studies as a resource for advancing a more public and democratic vision for higher education. Hopefully, it suggests theoretical tools for constructing new forms of collaboration among faculty, a broadening of the terms of teaching and learning, and new approaches to transdisciplinary research that address local, national, and international concerns. The potential that cultural studies has for developing forms of collaboration that cut across national boundaries is worth taking up.

The Worldly Space of Culture

Cultural studies offers a number of important contributions for scholars about what it might mean to articulate and reclaim higher education as a democratic public sphere. The contribution that cultural studies offers educators becomes meaningful to the degree that, as Amy Gutmann argues in another context, it "is committed to allocating educational authority in such a way as to provide its members with an education adequate to participating in democratic politics, to choosing among (a limited range of) good lives, and to sharing in the several sub-communities, such as families, that impart identity to the lives of its citizens."[28] A substantive and inclusive democracy provides the political and ethical referent for framing what we do as educators and the role we play in using particular forms of knowledge and practice to offer specific visions of the world, particularly as they legitimate for students a sense of place, identity, worth, and value. The tension between the reality and the promise of democracy gives meaning to the importance of connecting a pedagogy aimed at promoting a culture of questioning with a pedagogy that focuses on a politics of social responsibility and public intervention. Although cultural studies offers some valuable theoretical insights into these considerations, it does not go far enough.

Like any other academic field, cultural studies is marked by a number of weaknesses, which need to be addressed by even those educators who are drawn to some of its more critical assumptions.

First, there is a tendency in some cultural studies work to be simply deconstructive—that is, to refuse to ask larger questions about the nature and purpose of a democracy and what it means to connect matters of textuality to broader social considerations and projects—especially those that merge symbolic issues with considerations of power, connect culture with history, translate private concerns into public issues, and articulate matters of academic considerations with wider national and global forces. For instance, there is little understanding in some deconstructive approaches of how texts, language, and symbolic systems are historically situated and contextualized "within and by a complex set of social, political, economic and cultural forces."[29] As the exclusive focus of analysis, texts become hermetically sealed, removed from power relations—and so the terrain of struggle is reduced to a debate over the meanings that allegedly reside in such texts. Any workable form of cultural studies cannot insist exclusively on the primacy of signification over power, thereby reducing its purview to questions of meaning and texts. An obsession with texts on textuality often results in privileging literature and popular culture over history and politics. Within this discourse, material organizations and economic power disappear into some of the most irrelevant aspects of culture. A narrow focus on academic fads and cultural trivia take on the aura of serious social analyses and legitimate the most privatized forms of inquiry while simultaneously "obstructing the formulation of a publicly informed politics."[30]

Educators need to foreground the ways in which culture and power are related through an emphasis on what Stuart Hall calls "combining the study of symbolic forms and meanings with the study of power," or more specifically the "insertion of symbolic processes into societal contexts and their imbrication with power."[31] Douglas Kellner has also argued for years that any practical approach to cultural studies has to overcome the divide between political economy and text-based analyses of culture.[32]

But recognizing such a divide is not the same thing as overcoming it. Educators must anchor their own work, however diverse, in a radical project that seriously engages the promise of an unrealized democracy against its really existing forms. Central to such a project is to reject the assumption that theory can understand social problems without contesting them in public life. At the same time, it is crucial to any viable notion of cultural studies that it reclaim politics as an ongoing critique of domination and society as part of a larger search for justice. Any workable form of cultural politics needs a socially committed notion of injustice if we are to take seriously what it means to fight for the idea of the good society. We agree with Zygmunt Bauman when he argues that "If there is no room for the idea of wrong society, there is hardly much chance for the idea of good society to be born, let alone make waves."[33] Educators need to be more forceful and committed to linking their overall politics to modes of critique and collective action that address the presupposition that democratic societies are never too just, which means that a society must constantly nurture the possibilities for self-critique, collective agency, and forms of citizenship in which people play a fundamental role in shaping the material relations of power and ideological forces that bear down on their everyday lives. Moreover, the struggle over an inclusive and just democracy can take many forms, offers no political guarantees, and provides an important normative dimension to politics as a process of democratization that never ends. Such a project is based on the realization that a democracy that is open to exchange, question, and self-criticism never reaches the limits of justice—that is, it is never just enough, and is never finished. It is precisely the openended and normative nature of such a project that provides a common ground for educators and cultural studies theorists to share their diverse range of intellectual pursuits.

By linking higher education to the project of an as yet unrealized democracy, educators can move beyond those approaches to pedagogy that reify what is sometimes called either the teaching of the conflicts or the culture of questioning.[34] These positions fail to make clear the larger political, normative, and ideological considerations that inform their view of education, teaching, and visions of the future, assuming that education is predicated upon a particular view of the future that students should hold. Furthermore, both positions collapse the purpose and meaning of higher education, the role of educators as engaged scholars, and the possibility of pedagogy itself into a rather shortsighted and sometimes insular notion of method, particularly one that emphasizes argumentation and dialogue. Such approaches fail to raise broader questions about the social, economic, and political forces shaping higher education, as well as the fragility of democracy itself. They often neglect key questions about the relationship between higher education and unbridled market forces, or those forces that unequally value diverse groups of students within different relations of academic power, or what it might mean to make pedagogy a basis not merely for understanding but also for intervening in the larger world. Both the political nature of education and the political possibilities it might produce are often either dealt with in a trivial fashion or simply ignored.

Consequently, such approaches often reproduce a general misunderstanding of how teacher authority can be used to create the conditions for an education in democracy without necessarily falling into the trap of simply indoctrinating students.[35] For instance, Gerald Graff implies that any notion of critical pedagogy that is self-conscious about its politics and engages students in ways that offer them the possibility for becoming critical (what Lani Guinier calls the need to educate students "to participate in civic life, and to encourage graduates to give back to the community, which through taxes, made their education possible"[36]) either leaves students out of the conversation or presupposes too much, becoming a form of pedagogical tyranny. While Graff advocates strongly that teachers create the educational spaces that open up the possibility of questioning among students, he refuses to go further and connect pedagogical conditions that challenge how they think in the moment to the next step: encouraging them to think proactively about changing the world around them so as to extend its democratic possibilities. George Lipsitz criticizes academics like Graff who believe that connecting academic work to social change is a burden, arguing that they have been subconsciously educated to accept cynicism about the ability of ordinary people to change the conditions under which they live.[37] Matters of public scholarship that link learning to social change, however openended, seem to be utterly tainted if we are to believe Graff. The call for debate and argumentation seems particularly ineffectual politically when it comes to linking learning to preparing students not to "be vulnerable to racism, sexism, homophobia, or structural, class-based forms of injustice"[38]—or, for that matter, teaching students the knowledge and skills they will need to learn how to govern, contribute to the public good, address social ills, and fight for a democratic society. Teaching students how to argue, draw on their own experiences, or engage in rigorous dialogue says nothing about why they should engage in these actions in the first place. How the culture of argumentation and questioning relates to giving students the tools they need to fight oppressive forms of power, make the world a more meaningful and just place, and develop a sense of social responsibility is missing in Graff's work because this is part of the discourse of political education, which Graff simply equates with indoctrination or speaking to the converted.

Many educators like Graff are unable to distinguish between propaganda and critical pedagogy or what we later call the distinction between political and politicizing education.[39] Propaganda is used generally to misrepresent knowledge, promote biased views, or support a politics that presents itself as beyond question and critical engagement. While no pedagogical intervention should fall to the level of propaganda, a pedagogy that attempts to empower critical citizens can't and shouldn't avoid politics. Pedagogy must address the relationship between politics and agency, knowledge and power, subject positions and values, and learning and social change, while always being open to debate, resistance, and a culture of questioning. Otherwise, educators have no language for linking learning to forms of public scholarship that would enable students to consider the important relationship between democratic public life and education, politics, and civic agency.[40] Disabled by a depoliticizing, if not slavish allegiance to a teaching methodology, educators like Graff exhibit little interest in encouraging students to enter the sphere of the political and think about how they might participate in a democracy by taking what they learn "into new locations—a third grade classroom, a public library, a legislator's office, a park,"[41] or by taking up other collaborative projects that address the cultural politics of engaged citizenship in a democracy. In spite of Graff's pretense to neutrality, academics need to do more pedagogically than simply teach students how to be adept at forms of argumentation or how to draw upon their own experiences in the classroom. Students need to argue and question, but they also need much more from their educational experience. The pedagogy of argumentation in and of itself guarantees nothing but it is an essential step toward opening up the space of resistance toward authority, teaching students to think critically about the world around them, and recognizing interpretation and dialogue as a condition for social intervention and transformation in the service of an unrealized democratic order. As Amy Gutmann brilliantly argues, education is always political because it is connected to the acquisition of agency and the ability to struggle with relations of power, and thus is a precondition for creating informed and critical citizens. For Gutmann, educators need to link education to democracy and recognize pedagogy as an ethical and political practice tied to modes of authority in which the "democratic state recognizes the value of political education in predisposing [students] to accept those ways of life that are consistent with sharing the rights and responsibilities of citizenship in a democratic society."[42] Rather than claiming an alleged neutrality, this vision of pedagogy is directive and interventionist on the side of democratic education. We take this issue up again in a different context in chapter 6.

It is precisely this democratic project that affirms the critical function of education and refuses to narrow its goals and aspirations to methodological considerations. This is what makes critical education different from mere training. It is precisely the failure to connect higher education to its democratic functions and goals that provides rationales for a pedagogical philosophy that strips education of critical responsibility and democratic possibilities. Stanley Fish, the current Dean of Liberal Arts and Sciences at the University of Illinois at Chicago, for example, not only denies what he calls "the effectiveness of intellectual work,"[43] but also reduces pedagogy to little more than an administrative performance organized around "the selection of texts, the preparation of syllabus, the sequence of assignments and exams, the framing and grading of a term paper, and so on."[44] In an article in the *Chronicle of Higher Education*, Fish chastises academics with grandiose visions of inspiring students, supporting the position that you can have rigor in the university but not educational practices that offer students the possibility of becoming social agents actively and critically involved in public life.[45] While we agree that providing students with the knowledge and skills they need to struggle to strengthen democracy does not guarantee that they will do so, it does seem imperative morally and politically to at least afford them the knowledge and skills that enable them, as Edward Said puts it, to uncover, elucidate, "challenge, and defeat both an imposed silence and the normalized quiet of unseen power, wherever and whenever possible."[46] At the very least, we believe that such educational efforts are a precondition, rather than a guarantee, to challenge the currently fashionable neoliberal view that there are no alternatives to the way society is organized or that, as Margaret Thatcher famously put it, "there is no such thing as society."[47]

Fish argues against the practical value of pedagogical practices that provide students with both the knowledge and capacities to contest power or the range of choices affording economic, political, cultural, social, and intellectual development. He seems to be oddly incapable of grasping the relationship between education and the production of particular forms of social agency, just as he denies, even in the weakest sense, Geoffrey Hartman's insight that "There is a link between epistemology and morality: between how we get to know what we know, and the moral life we aspire to lead."[48] Fish's zest for efficiency in educational outcomes and his predictable privileging of disciplinary rigor over ethical deliberations reveals both a certain disdain for the role the university might play as a democratic public sphere and a contempt for those educators and students who feel a deeper civic responsibility for their pedagogical actions.[49] Fish's fashionable cynicism toward the teaching of democratic values, which he dismisses vulgarly as self-help therapy, might explain the dislike he displays in one particular article not only toward the realm of the personal but also toward a student who thanks him years after having taken his course.[50] Of course, nobody can accuse Fish of hypocrisy, because he named his piece "Aim Low."[51]

It is precisely in this context that Raymond Williams provides an important insight for academics by insisting that cultural studies theorists understand and engage the importance of the project that drives their work. Following Williams, we believe that academics need to situate their work within the university as part of a broader project of democratization that provides knowledge, classroom experiences, and pedagogical engagements that give students the opportunity to embrace and defend democratic values. Again, as Gutmann observes, learning the moral values that distinguish a democratic society does not happen under pedagogical practices that primarily view themselves as neutral methodologies rather than political and moral practices. Learning that bigotry, racism, or the sexual exploitation of children is bad or morally indefensible, does not take place by simply offering it up as one opinion among many or as one among many conceptions of the good life to be debated as part of a culture of conflict. Bigotry as a threat to democracy has to be identified as such, and then compelling reasons have to be provided to justify the argument, but always within a spirit of critical dialogue, historical engagement, and democratic values.

Second, cultural studies is still largely an academic discourse and as such is often too far removed from other cultural and political sites where learning takes place. In order to create a public discourse of any importance, cultural studies theorists will have to focus their work on the immediacy of more public problems that are relevant to important social issues. These might include the destruction of the biosphere, the war against youth, the increasing corporate control of media, the widespread attack by corporate culture on public schools, the ongoing attack on the welfare system, the increasing rates of incarceration of people of color, the widening gap between the rich and the poor, the fiscal crisis of the state, the decline of civic organizations and the depoliticization of politics, the increasing global spread of war, or the dangerous growth of the prison-industrial complex. To effectively engage such concerns, cultural studies theorists need to write for a variety of audiences rather than simply for a narrow group of specialized intellectuals. Such writing needs to become public by crossing over into sites available to more general audiences and by using a language that is clear but not theoretically simplistic, scholarly but not dull or obtuse. By engaging public means of expression including the lecture circuit, radio, the internet, interviews, alternative magazines, and the church pulpit, to name only a few, academic intellectuals can expand their audiences. Of course, there is the need, as Pierre Bourdieu reminds us, of using these public spheres not to reproduce sound bytes or spectacles but to make decisive arguments.

Third, educators and cultural studies theorists should be more willing and specific about what it means to work collectively with other academics and researchers through a vast array of networks across a number of public spheres on important domestic problems. This points to the necessity for academics to engage in research that addresses vital social issues and share their intellectual resources with community activists, religious groups, citizens, and young people who are actively involved in vital social, economic, and political issues. For instance, academics could work with local groups in Florida in trying to prevent the Florida Panhandle from being taken over by land developers; join with local and national groups in fighting the commercialization and privatization of public schools; work with the various groups agitating for global justice or those activists battling the ongoing destruction of state provisions and vital public resources in crisis states such as California, Texas, and Oregon. Academics could address a number of youth issues, playing a vital role in challenging the assault on public education now being waged by Bush and his fellow travellers. This might include a range of possible collective actions such as fighting against tax cuts for the rich in an effort to force the federal government to provide the much-needed financial backing to rebuild the crumbling infrastructure of public schools and health services. Academics could use their collective knowledge and resources to challenge those educational policies organized around drill and rote learning and the crazed obsession with standardized testing that is resulting in massive numbers of students being pushed out of schools, administrators lying about dropout rates, and teachers being reduced to test preparers—a position that both deskills and disempowers teachers.[52] The key here is the need for academics to become engaged not as professional experts but as allies with particular resources that expand the possibility for listening to others and working to get people to participate in public life around important issues. This is especially relevant in getting young people to participate in the realms of politics and critical education, as seen in Adam Fletcher's work with The Freechild Project or the pioneering work of the Urban Debate League, which gets inner city kids actively involved as critical agents by organizing debates for those students who are

marginalized and dispossessed.[53] Other projects combining politics
and public pedagogy can be found in Sut Jhally's Media Education
Foundation, which produces critical films and videos on a range of
crucial issues, providing an important site for an alternative form
of public pedagogy. Engaged academics could create their own
think-tanks, produce public intellectuals actively engaged in creating alternative radio, progressive web sites such as www.
commondreams.org, television programming comparable to what
Doug Kellner and Robert McChesney have been doing for years, or
join with various youth organizations such as Youth Led Art for
Social Change and Youth Fighting Racism. This suggests that academics make connections to public life in their research and teaching. It also points to the possibility for academics to reach out and
become citizen-scholars involved in mobilizing organizations that
promote active engagement in democratic public life. On a global
level, academics are becoming increasingly active in addressing the
ethical and political challenges of globalization. As capital, finance,
trade, and culture become extraterritorial and removed from traditional political constraints, it becomes all the more pressing that
global networks and political organizations be put into play to provide an effective response. Engaging in intellectual practices that
offer the possibility of alliances and new forms of solidarity among
cultural workers such as artists, writers, journalists, academics,
and others who engage in forms of public pedagogy grounded in
a democratic project represents a small but important step
in addressing the massive and unprecedented reach of global capitalism. The issue here is that academics don't need to reproduce
the professional managerial/expert class. On the contrary, they
need to unlearn the vanguardist privileging that comes with expert
knowledge and relearn what it means to use one's knowledge as
part of a broader movement to create organizations that encourage
rather than shut down citizen participation in political culture
and life.[54]

Fourth, educators need to make visible their own subjective
involvement in what they teach, how they shape classroom social
relations, and how they defend their positions within institutions
that often legitimate education processes based on ideological
privileges and political exclusions. Making one's authority and

classroom work the subject of critical analysis with students is important, but such a task must be taken up in terms that move beyond the rhetoric of method, psychology, or individualizing interests. As we previously argued, pedagogy can better be addressed as a moral and political discourse in which students are able to connect learning to social change, scholarship to commitment, and classroom knowledge to public life. Such a pedagogical task points to the necessity for educators and cultural theorists to define intellectual practice "as part of an intricate web of morality, rigor and responsibility"[55] that enables them to speak with conviction, expand the concept of the political, enter the public sphere in order to address important social problems, and demonstrate alternative models for bridging the gap between higher education and society. Social critique is inextricably linked to self-critique not only because it provides a space for students and others to question authority, but also because it places values and ideology at the center of one's actions. Authority under such circumstances is not legitimated by extra-social appeals to god, biology, history, or the "hidden hand" of the market, but to the arguments, rules, knowledge, and experiences one musters in the world of real human beings. That is the complicated world of history, politics, values, and power.

Making authority accountable in the first instance means being self-critical about how one justifies one's relationship to official power and to the exercise of power itself—particularly as such power is exercised through the force of classroom practices and relationships. One useful approach is for educators to think through the distinction between a politicizing pedagogy, which insists wrongly that students think as we do, and a political pedagogy that teaches students by example the importance of taking up critical positions without becoming dogmatic or intractable. Political pedagogy connects understanding with social responsibility and seeks to educate students not only to critically engage the world but also be responsible enough to fight for those political and economic conditions that make democracy possible. Such a pedagogy affirms the experience of the social and the obligations it evokes regarding questions of responsibility and social transformation. It does so by opening up for students important questions about power, knowledge, and equality, and what it might mean for them to work to overcome those social relations of oppression that make living unbearable for those who are poor, hungry, unemployed, refused adequate social services, and viewed under the aegis of neoliberalism as largely disposable. Central here is the importance for cultural studies educators to encourage students to reflect about what it would mean for them to connect knowledge and criticism to becoming social actors, buttressed by a profound desire to overcome injustice and a spirited commitment to social agency. Political education teaches students to take risks and challenge those with power, and encourages them to be conscious of how power is used in the classroom. *Political* education proposes that the role of the public intellectual is as Edward Said argues "not to consolidate authority, but to understand, interpret, and question it," and that teachers and students should temper any reverence for authority with a sense of critical awareness and an acute willingness to hold it accountable for its consequences.[56] Moreover, it situates education not within the imperatives of specialization and professionalization, but within a project designed to expand the possibilities of democracy by linking education to modes of political agency that promote critical citizenship and engage the ethical imperative to alleviate human suffering. On the other hand, *politicizing* education silences in the name of orthodoxy and imposes itself on students while undermining dialogue, deliberation, and critical engagement. Politicizing education is often grounded in a combination of selfrighteousness and ideological purity that silences students as it imposes "correct" positions. Authority in this perspective rarely opens itself to selfcriticism, or for that matter to any criticism, especially from students. Politicizing education cannot decipher the distinction between critical teaching and indoctrination because its advocates have no sense of the difference between encouraging human agency and social responsibility and molding students according to the imperatives of an unquestioned ideological position. Politicizing education is more concerned with the sacred than the secular, more about training than educating, and it harbors a great dislike for critical dialogue and a culture of questioning.

Finally, if educators, like many cultural studies theorists, are truly concerned about how culture operates as a crucial site of power in the modern world, they will have to take more seriously how pedagogy functions on local and global levels to secure and challenge the ways in which power is deployed, affirmed, and resisted within and outside traditional discourses and cultural spheres. Pedagogy thus becomes an important theoretical tool for understanding the institutional conditions that constrain the production of knowledge, learning, and academic labor itself. Pedagogy also provides a discourse for engaging and challenging social hierarchies, identities, and ideologies that transcend local and national borders. In addition, pedagogy offers a discourse of possibility, a way of providing students with the opportunity to link meaning to commitment and understanding to social transformation—and to do so in the interest of the greatest possible justice. Unlike traditional vanguardist or elitist notions of the intellectual, cultural studies should embrace the idea that the vocation of intellectuals be rooted in pedagogical and political work tempered by humility, a moral focus on suffering, and the need to produce alternative visions and policies that go beyond a language of critique. We now want to shift our frame a bit in order to focus on the implications of the concerns we have addressed thus far and how they might be connected to developing an academic agenda for teachers as public intellectuals in higher education, particularly at a time when neoliberal agendas increasingly guide social policy and threaten higher education as a sphere for critical teaching and learning.

The Responsibility of Intellectuals and the Politics of Education

> We are witnessing the mutation of a new, global body politic, and if we intellectuals are to have any potency as part of its thinking organ, it will be in discourses that refuse to separate academic life from political life, and that inform not just national opinion, but a global public debate.[57]

In opposition to the commodification, privatization, and commercialization of everything educational, educators need to define higher education as a resource vital to the democratic and civic life of the nation. The challenge is thus for academics, cultural workers, students, and labor organizers to join together and oppose the transformation of higher education into a commercial sphere, to resist what Bill Readings has called a consumer-oriented corporation more concerned about accounting than accountability.[58] As Zygmunt Bauman reminds us, schools are one of the few public spaces left where students can learn the "skills for citizen participation and effective political action. And where there is no [such] institutions, there is no 'citizenship' either."[59] Higher education may be one of the few sites available in which students can learn about the limits of commercial values, the skills of social citizenship, and how to enlarge the possibilities of collective agency and democratic life. Defending education as a vital public sphere and public good rather than merely a private good is necessary to develop and nourish the proper balance between democratic public spheres and commercial power, and between identities founded on democratic principles and identities steeped in forms of competitive, self-interested individualism. This view suggests that higher education (as well as public education) be defended through intellectual work that self-consciously recalls the stress between the democratic imperatives and possibilities of public institutions and their everyday realization within a society dominated by market principles. If colleges and universities are to remain sites of critical thinking, collective work, and social struggle, public intellectuals need to expand their meaning and purpose. As we have stressed repeatedly, academics, teachers, students, parents, community activists, and other socially concerned groups must provide the first line of defense of higher education as a resource vital to the moral life of the nation, open to people and communities whose resources, knowledge, and skills have often been viewed as marginal. Educators and cultural studies theorists need to develop a more inclusive vocabulary for aligning politics to the tasks of civic leadership. In part, this means providing the language, knowledge, and democratic social relations for students to engage in the "art of translating individual problems into public issues, and common interests into individual rights and duties."[60] Leadership demands a politics and pedagogy that refuses to separate individual problems and experience from public issues and social considerations. Within such a perspective, leadership displaces cynicism with hope, challenges the neoliberal notion that there are no alternative visions of a better society, and develops a pedagogy of commitment that puts into place modes of literacy in which competency and interpretation provide the basis for actually intervening in the world. Leadership invokes the demand to make the pedagogical more political by linking critical thought to collective action, human agency to social responsibility, and knowledge and power to a profound impatience with a status quo founded upon deep inequalities and injustices.

One of the most crucial challenges that educators and cultural studies advocates face is rejecting the neoliberal collapse of the public into the private—the rendering of all social problems as personal. The neoliberal obsession with the private not only furthers a market-based politics that reduces all relationships to the exchange of money and the accumulation of capital; it also depoliticizes politics itself and reduces public activity to the realm of utterly privatized practices and utopias, limiting citizenship to the act of buying and purchasing goods. Political solidarity, social agency, and collective resistance disappear into the murky waters of a biopolitics in which the pursuit of private pleasures and ready-made individual choices are organized on the basis of marketplace aims and desires, which cancel out all social responsibility, commitment, and action.[61] The current challenge intellectuals face is to reclaim the language of the social, agency, solidarity, democracy, and public life as the basis for rethinking how to create a new kind of politics, political agency, and collective struggle. Equally important is the challenge of providing a language that addresses what it means to both theorize a notion of democracy suitable to a new global public sphere and a politics capable of new forms of shared solidarity.

Positing new forms of social citizenship and civic education can have a profound effect on people's everyday lives and struggles. Academics bear an enormous responsibility in opposing neoliberalism—the most dangerous ideology of our time—by bringing democratic political culture back to life. New locations of struggle, vocabularies, subject positions, and modes of exchange will have to be created that allow people in a wide variety of public spheres to speak and act together. Such an effort also requires a language that is not simply negative, a critical discourse steeped in a sense of utopian longing in which people gain some control over the commanding forces shaping their lives, question what it is they have become within existing institutional and social formations, and "give some thought to their experiences so that they can transform their relations of subordination and oppression"[62] while not falling into a new dogmatism. One element of this struggle could take the form of resisting attacks on existing public spheres such as schools while creating new spaces in clubs, neighborhoods, bookstores, trade unions, alternative media sites, and other places where dialogue and critical exchanges become possible. But, as Jo Ellen Green Kaiser argues, contemporary society lacks more than the material spaces, "which democracy needs to flourish. . . . American culture [also] lacks the time for democracy to grow and flourish. Citizens simply don't have the . . . time to formulate, enunciate, and act upon their highest ideals."[63] This suggests that educators need to work with students and others in order to win back their public voice and gain some control over their time in rigid administrative and hierarchical structures, especially in the university and the workplace. By linking the crisis of democracy and citizenship to the crisis of time, academics could connect struggles against the increasing economic conditions under which part-time faculty and students work to those economic, political, and cultural conditions outside of the university that urge people to work more with less time. At stake here is addressing those diverse conditions that extend from inadequate public transportation to insufficient health care to extended work hours that make time a deprivation for so many individuals and in doing so prevents them from conducting the work of citizenship in the larger society.[64] At the same time, challenging neoliberalism means fighting against the ongoing reconfiguration of the state into the role of an enlarged police precinct designed to repress dissent, regulate immigrant populations, incarcerate youth who are considered disposable, and safeguard the interests of global investors. As governments globally give up their role of providing social safety nets, a living wage, and regulating the excesses of corporate greed, capital escapes beyond the reach of democratic control while marginalized groups are left to their own meager resources to survive. Under such circumstances, it becomes difficult to create alternative public spheres that enable people to become effective agents of change. Under neoliberalism's reign of terror, public issues collapse into privatized discourses and a culture of personal confessions, survivor stories, and celebrity sightings sets the stage for a depoliticized public life in which citizenship is replaced by consumerism, and a government of corporate-friendly parties replaces a government of citizens.

Against neoliberalism, educators, cultural studies theorists, students, and activists face the task of "showing how the space of the possible is larger than the one assigned—that something else is possible, but not that everything is possible."[65] This points to the hard work of providing a language of resistance and possibility, a language that embraces a militant utopianism while constantly being attentive to those forces that seek to turn such hope into a new slogan or punish and dismiss those who dare look beyond the horizon of the given. Such a language must address, as Dick Hebdige observes, how different futures can be imagined, and what strategies can be used to "open up or close down particular lines of possibility."[66] Educated hope, in this context, becomes the affective and intellectual precondition for individual and social struggle. By anticipating a better world, hopeful education combines reason with a gritty sense of the mutually constitutive relationship between limits and possibilities. There is a lot of talk among social theorists about the death of politics and the inability of human beings to imagine a more equitable and just world in order to make it better. We would expect that, of all groups, educators would be the most vocal and militant in challenging this assumption by making it clear that at the heart of any form of inclusive democracy is the assumption that learning should be used to expand the public good, create a culture of questioning, and promote democratic social change. Individual and social agency becomes meaningful as part of the willingness to imagine otherwise "in order to help us find our way to a more human future."[67] Under such circumstances, knowledge can be used for amplifying human freedom and promoting social justice, and not simply for creating profits. The diverse but connected fields of cultural studies and critical pedagogy offer some insights for addressing these issues, and we would do well to learn as much as possible from them in order to expand the meaning of the political and revitalize the pedagogical possibilities of cultural politics and democratic struggles. The late Pierre Bourdieu has argued that intellectuals need to create new ways for doing politics by investing in political struggles through a relentless critique of the abuses of authority and power. Bourdieu wanted scholars to use their skills and knowledge to break out of the microcosm of academia, combine scholarship with commitment, and "enter into sustained and vigorous exchange with the outside world (especially with unions, grassroots organizations, and issue-oriented activist groups) instead of being content with waging the 'political' battles, at once intimate and ultimately, and always a bit unreal, of the scholastic universe."[68]

At a time when our civil liberties are being destroyed through laws such as the USA PATRIOT Act, and public institutions and goods all over the globe are under assault by a rapacious global capitalism, Bourdieu is right in emphasizing an urgent need for the most militant forms of political opposition on the part of academics, as well as new modes of resistance and collective struggle buttressed by rigorous intellectual work, social responsibility, and political courage. Intellectuals need to recognize the ever-fashionable display of rhetorical cleverness as a form of "disguised decadence."[69] They need to view democracy as a site of struggle, that demands more than irony or the deconstructing of texts. While critical analysis is crucial to any substantive notion of politics, it is the contradiction between the reality and the promise of democracy that provides the conditions for collective resistance and struggle. We have seen glimpses of such a promise among those brave students, environmental activists, and workers who have demonstrated in Seattle, Genoa, Prague, New York, and Toronto against the World Trade Organization (WTO). As public intellectuals, academics can learn from such struggles by turning the university and public schools into vibrant critical sites of learning and unconditional sites of pedagogical and political resistance. The power of the dominant order does not merely reside in the economic realm or in material relations of power, but also in the realm of ideas and culture. This is why as we stress throughout this book, intellectuals must take sides, speak out, and engage in the hard work of debunking corporate culture's assault on teaching and learning. They must orient their teaching toward social change, connect learning to public life, link knowledge to the operations of power, and allow issues of human rights and crimes against humanity in their diverse forms to occupy a space of critical and open discussion in the class-room. It also means stepping out of the classroom and working with others to create public spaces where it becomes possible to not only "shift the way people think about the moment, but potentially to energize them to do something differently in that moment," to link one's critical imagination with the possibility of activism in the public sphere.[70] This is, of course, a small step, but if we do not want to repeat the present in the future or, even worse, become complicitous in the dominant exercise of power, it is time for edu-cators to mobilize collectively by breaking down the illusion of unanimity that dominant power propagates while working tire-lessly to reclaim the promises of a truly global, democratic future.

Part II

Higher Education and the Politics of Race

Chapter 4

Race, Rhetoric, and the Contest over Civic Education

The Liberal Arts in a Neoliberal Age

It is one of the more revealing paradoxes in contemporary liberal arts education that recent, cutting-edge discourses proffered in the service of democratic renewal—discourses frequently excoriated as trendy, postmodern, or ultra-radical by academics and the popular press alike—share, in many ways, the assumptions of some of the oldest theoretical justifications for higher education in America. Primarily concerned with reasserting the university's role in producing a literate and critical citizenry, recent progressive work in rhetorical and cultural theory has focused on the dynamic interconnections among the study of rhetoric and composition, the practice of democratic citizenship, and the politics of race.[1] In doing so, such work speaks to the necessity of an educational discourse steeped in democratic principles at a time when neoliberal agendas redefine public goods such as schooling as private interests, and in doing so suggest that "we have no choice but to adapt both our hopes and our abilities to the new global market."[2]

For those unfamiliar with the history of American universities or the social foundations of education, the relationship among terms such as *rhetoric, pedagogy, democracy, ethics*, and *race* are not immediately apparent. Nor will this particular combination of topics fall easily on the ears of those in academia who insist that education, even civic education, can somehow be abstracted from broader questions of politics in a multiracial and multiethnic society. Although recent theoretical work points to the necessity of taking up a fundamental commitment to democracy as an ongoing educational and ethical project within the field of rhetoric and composition, and liberal arts education in general, such a call is not entirely new. Dedication to education for democracy, for example, can be traced as far back as the radical educational work of Thomas Jefferson, the Enlightenment philosopher and statesman who was one of the first to put forth a multitiered plan for free and universal public education as the primary means for safeguarding a young and fragile democratic nation. Of course, the Jeffersonian legacy is also central to any understanding of the nation's most vexing contradiction—a historical commitment to universal citizenship and free, public education that simultaneously excluded nonwhite races and women. We are not suggesting, however, that current progressive work is merely a recuperation of a forgotten rhetorical model of university education, but rather that it is an attempt to locate such work within a tradition of thought about the relationship between higher education and the practice of citizenship, while at the same time demonstrating where that work departs from tradition to engage its most critical theoretical weaknesses and exclusions.

Recently, there has been an odd convergence of rhetorics deployed by academics from left to right in the contest over the future of liberal arts education. The language of curricular reform has expanded. Whereas "culture," or the more specific "canon," was *the* contested terrain in the academy a decade ago, battlelines are now being drawn around notions of "citizenship" and "civic education" as well. The broadening of this theater of struggle is not necessarily a negative turn of events; it may even produce more rather than less latitude for negotiation among generally opposed ideological positions in the humanities. In contrast to the go-nowhere debates over culture—the Matthew Arnold-or-bust idiom of the right versus an often essentialized identity politics on the left—civic education offers a language of social responsibility and social change often lost in the allegiances to the individual cultivation of pure taste or narrowly defined group solidarities. Certainly this has been the case at dozens of schools such as Berkeley, University of Wisconsin, Harvard, Cornell, and George Mason University, where student and faculty protests against the growing corporate influence on research and curricular requirements have recently erupted. As Kevin Avruch, a professor of anthropology at GMU, noted, such restructuring has "actually united professors on the left and right."[3] Avruch explains that although the faculty at GMU are often characterized as "overly liberal," they discovered that they had at least one thing in common with their colleagues on the Right: "we share a nineteenth-century view that our job is to educate well-rounded citizens."[4] Thus, the rhetoric of civic education also provides a shared language informed by democratic—rather than market—traditions to fight the ongoing vocationalization and corporatization of higher education.[5] At the same time, citizenship, like culture, is not a stable referent. As often as appeals are made to the education of future generations of citizens in a variety of academic venues, there is shockingly little attention given to the different ways in which citizenship as an ideal and as a set of practices is defined and negotiated both currently and historically. As Judith Skhlar aptly notes, "there is no concept more central in politics than citizenship, and none more variable in history, or contested in theory."[6]

Hence, our continued reliance on war metaphors is not accidental; we use them to dramatize our efforts to shift the debate over liberal arts education, in Chantal Mouffe's terms, from the realm of antagonism to one of agonism. If, as Mouffe explains, an antagonism defines a "relation between enemies" in which each group wants to destroy the other, then agonism marks a relation among "adversaries" who struggle "in order to establish a different hegemony."[7] Our goal, correspondingly, is not to wage a polemical war, the point of which is a simple dismissal of conceptions of citizenship and civic education other than our own; rather, it is an attempt to bring historical evidence to bear on an evaluation of different articulations of citizenship and corresponding forms of education. The purposes of this chapter, then, are threefold: first, it seeks to reaffirm critical citizenship as a core value and the centrality of civic education to democratic public life at a time when "visionary reform" has led to the corporatization of the university and capitalism has become synonymous with democracy itself. Second, it maps the history of various definitions of citizen—liberal, republican, and ascriptive Americanist[8]—and the forms of education proper to their development in an effort to establish the centrality of race and rhetoric to current debates over the future of liberal arts education. Finally, it examines the necessary and historical linkages between educational theory and curricular development, the practice of citizenship, and the politics of race. Our interest in exploring the various definitions of citizenship and civic education at work in contemporary professional conversations is not to establish the objective equality of all positions. Although disparate understandings of these key notions demand due consideration, we will nonetheless provide a very specific interpretation of the social values different theoretical positions represent as we defend our own project as part of a broader effort to connect learning to the production of democratic values and the imperative of emancipatory social change.

Before we examine the current controversy over the role of a liberal arts education in the production of good citizens, it is necessary to first address the various ways in which the concept of the "good citizen" has been defined over time. Hence, in what follows, we will map different conceptions of American civic identity, indicating when, historically, each enjoyed a period of relative hegemony. In doing so, we will analyze how shifts in dominant notions of citizenship in the last decades of the nineteenth century correspond to significant changes in college curricula about the same time—changes that dramatically altered the nature and purpose of higher education for the next century. Our hope is to establish a relevant historical context for, and so a richer assessment of, the contemporary debates over these issues. Of course, it is impossible to render this extensive history in any nuanced or complete way here, so we offer only passing apologies for the necessary simplification involved.

Conflicting Visions of American Citizenship: Liberal, Republican, and Ascriptive

Since Alexis de Tocqueville's 1840 classic, *Democracy in America*, the traditions of American political philosophy have held that citizenship is not determined by birth or inherited traits, but rather by sworn adherence to a set of political ideals, principles, and hopes that comprise liberal democracy. According to the liberal perspective, to be an American citizen, a person did not have to be of any particular national, linguistic, religious, or ethnic background (though racialized minorities were not recognized as citizens until the twentieth century). All one had to do was to pledge allegiance to a political ideology centered on the abstract ideals of liberty, equality, and freedom, largely derived from the seventeenth-century philosopher John Locke. Conceived in opposition to the oppressive hierarchies of traditional or feudal societies—societies dominated by a monarchy, aristocracy, or the church—liberalism has always maintained "a contractual and competitive rather than ascriptive idea of social order."[9] Rather than accept the rigid "social hierarchies characteristic of conservative social philosophies," liberalism has always been on the side of "change, dynamism, growth, mobility, accumulation and competition."[10] Accordingly, liberalism has tended to stand for a commitment to individualism, upholding the moral, political, and legal claims of the individual over and against those of the collective. It vouchsafes universal rights applicable to all humans or rational agents, the force of reason, and so rational reform. Liberalism privileges equality and religious toleration rather than repressive medieval religious and intellectual orthodoxies, hence the defense of pluralism, the division of church and state, and progress through the promotion of commerce and the sciences.[11] Citizenship in a liberal polity, then, is not a function of birthright or inheritance, but the right of any energetic individual who has achieved social standing and success through the pursuit of his own interests.[12] Thus, as Philip Gleason argues, "the universalist ideological character of American nationality meant that it was open to anyone who willed to become an American."[13] From Tocqueville to Louis Hartz's 1955 classic, *The Liberal Tradition in America*, the Lockean liberal foundation of American politics enjoyed an uncontested hegemony.[14]

Beginning in the late 1960s, however, the received understanding of American political culture as overwhelmingly liberal democratic was significantly challenged in at least three ways. Following the lead of Bernard Bailyn and his groundbreaking 1967 publication,

The Ideological Origins of the American Revolution, a number of historians such as Gordon Wood, John Pocock, Lance Banning, and others have claimed that American political philosophy was shaped by traditions of republicanism that were different from, and in significant ways opposed to, the liberalism of John Locke. According to Pocock, the origins of civic republicanism extend back to the works of Aristotle and Cicero, but it is in the fifteenth-century Florence of Machiavelli that such traditions find their apotheosis and go on to influence American political thought. In contrast to liberalism's conception of liberty as freedom from state interference in individual private pursuits, the common feature of the diverse strains of republican thought was "an emphasis on achieving institutions and practices that make collective self-governance in the pursuit of a common good possible for the community as a whole."[15]

Against the liberal concern for the individual's universal rights and freedoms, the second critique ran in a similar vein to the first. Communitarian political theorists like Michael Sandel and Alasdair MacIntyre acknowledged the dominance of liberal philosophy in American thought, but they also argued that the liberal conception of the individual was an entirely atomistic one, leaving no room for a theory of political community or a notion of public good. In other words, because liberalism held the individual to be "naturally" driven by power, competition, self-interest, and security, it follows that the liberal concept of the good society was one in which individuals could pursue their private affairs with the least interference. The few constraints society imposed were necessary to ensure the equal protection of all in the common pursuit of their self-interests, to prevent individuals from destroying one another in the Hobbesian "war of all against all." Hence, liberalism had no way to engage the desire for, or necessity of, meaningful collective political life or pride in origin, let alone accommodate such notions as public-mindedness, civic duty, or active political participation in a community of equals.

In contrast to the liberal tension between the individual and the state, republican thought favors free popular government, requiring citizens to actively participate in their own self-rule. Although liberalism has contributed the notion of universal citizenship to American political thought, it has also reduced it to a mere legal status.[16] Conversely, civic republicanism holds citizenship to be an ongoing activity or a practice. Moreover, "civic republicanism," as Adrian Oldfield has argued, "recognizes that, unsupported, individuals cannot be expected to engage in the practice [of self-governance]. This means more than that individuals need empowering and need to be afforded opportunities to perform the duties of the practice: it means, further, that they have to be provided with sufficient motivation."[17] For Oldfield, the motivations for active political citizenship include the capacity to attain a degree of moral and political autonomy that a liberal rights-based citizenship cannot vouchsafe. Civic republicanism also maintains that direct participation in the political life of the nation creates the conditions for the highest form of moral and intellectual growth. In addition to full political participation, republicanism also requires that citizens acknowledge the goals of the political community and the needs of individuals as one and the same—hence Montesquieu's argument that citizens in a classical republic had to be raised "like a single family."[18] Identification with one's political community is achieved through "a pervasive civic education in patriotism reinforced by frequent public rites and ceremonies, censorship of dissenting ideas, preservation of a single religion if possible, limits on divisive and privatizing economic pursuits, and strict restraints on the addition of aliens to the citizenry."[19] Thus, a successful republic is characterized by considerable social homogeneity and must be composed of a relatively small number of citizens.

According to Rogers Smith, such demands have no small role to play in justifying a wide range of political exclusions and inequalities. "The demand for homogeneity," Smith concludes,

> could be used to defend numerous ethnocentric impulses including citizenship laws that discriminated on the basis of race, sex, religion and national origins. The second requirement helped generate and maintain America's commitment to federalism, to state and local autonomy—a commitment often used to justify national acquiescence in local inequalities.[20]

Finally, scholars such as Smith and Judith Skhlar have recently extended the liberal critique by taking up the question of American civic identity from the perspective of historically excluded groups—women and minorities of color. Skhlar has demonstrated how institutionalized forms of servitude were not anomalous to but absolutely constitutive of a modern popular representative republic dedicated to liberty and freedom. For Skhlar, "The equality of political rights, which is the first mark of American citizenship, was proclaimed in the accepted presence of its absolute denial. Its second mark, the overt rejection of hereditary privileges, was no easier to achieve in practice for the same reason."[21] Similarly, Smith set out to assess the civic republican critiques through an investigation of American citizenship laws, which both defined what citizenship was and who was capable of achieving it. The upshot was a 700-page tome entitled *Civic Ideals: Conflicting Visions of Citizenship in U.S. History* and a fundamental redefinition of American political culture. In short, Smith contends that though many liberal and republican elements were visible, much of the history of American citizenship laws did not fit the liberalism of Montesquieu and Hartz or the republicanism of Pocock and MacIntyre. Smith argues:

> Rather than stressing the protection of individual rights for all in liberal fashion, or participation in common civic institutions in republican fashion, American law had been shot through with forms of second-class citizenship, denying personal liberties and opportunities for political participation to most of the adult population on the basis of race, ethnicity, gender, and even religion . . . many of the restrictions on immigration, naturalization, and equal citizenship seemed to express views of American civic identity that did not feature either individual rights or membership in a republic. They manifested passionate beliefs that America was by rights a white nation, a Protestant nation, a nation in which true Americans were native-born men with Anglo-Saxon ancestors.[22]

Accordingly, Smith boldly identifies yet another tradition in American political thought in addition to liberalism and civic republicanism, the "ascriptive tradition of Americanism." From the dawn of the republic, Smith explains, many Americans defined citizenship not in terms of personal liberties or popular self-governance but rather in terms of

> a whole array of cultural origins and customs—with northern
> European, if not English ancestry, with Christianity, especially
> dissenting Protestantism, and its message for the world; with the
> white race, with patriarchal familial leadership and female domes-
> ticity; and with all the economic and social arrangements that came
> to be seen as the true, traditional "American" way of life.[23]

According to Smith, ascriptive Americanism, or the identification of American nationality with a particular ethnocultural identity, became a full-fledged civic ideology by the late nineteenth century, spurred by such events as the growth of racial science, the alarm over mass European immigration, and the desire to dismantle those social policies associated with Reconstruction. And formal institutions of education, as widely noted, have been one of the primary vehicles for producing a largely assimilationist version of American citizenship.[24]

It is important to note that Smith's multiple traditions thesis is not an attempt to shift responsibility for the vast inequalities of American life onto its ascriptive traditions, exonerating liberal and republican values and institutions. To be sure, Matthew Frye Jacobson and David Theo Goldberg have insightfully demon-strated how republican and liberal traditions have been complici-tous with racialized ideologies and exclusions. Jacobson argues, for example, that citizenship was a racially inscribed concept from the start of the new nation; "political identity was rendered racial identity"—thus establishing, at least implicitly, a European politi-cal order in the New World.[25] Though the majority of scholarly opinion has decried a democracy built on both gender and racial exclusion as both a profound hypocrisy and a betrayal of its most sacred principles, Jacobson asserts that racial and gendered exclu-sions cannot be understood as a mere inconsistency in an otherwise liberal political philosophy; on the contrary, racialism is insepara-ble from, and in fact constitutive of, the ideology of republicanism. Both the tenets of classical republicanism and the racist practices that normalized the equation of whiteness with citizenship have deep roots in Enlightenment thought.

According to Jacobson, the Enlightenment experiment in democratic forms of government demanded "a polity disciplined, virtuous, self-sacrificing, productive, far-seeing, and wise—traits that were all racially inscribed in eighteenth-century Euro-American thought."[26] In short, the shift from monarchic power to democratic power demanded of its participants a remarkable degree of "self-possession"—a condition that was already denied literally to Africans in bondage and figuratively to both "savage" or "non-white" peoples, as well as women, who were said to be lacking in reason, dispassionate judgment, and overall "fitness for self-government."[27] And republicanism, with its emphasis on the common good, community, and self-sacrifice, also demanded from "the people" an extraordinary moral character. At a time when the Anglo-Saxon was hailed as a paragon of political genius, reflection, and restraint, Jacobson wryly notes that a definition of the word Negro in a Philadelphia encyclopedia could include "idleness, treachery, revenge, debauchery, nastiness, and intemperance."[28]

Similarly, Goldberg has eloquently elaborated on liberalism as the preeminent modern—and modernizing—ideology and its central paradox: as modernity commits itself to the idealized principles of liberty, equality, and fraternity, as it increasingly insists upon the moral irrelevance of race, there is a proliferation of racial identities and sets of exclusions that they rationalize and sustain. "The more abstract modernity's universal identity," he explains, "the more it has to be insisted upon, the more it needs to be *imposed*. The more ideologically hegemonic liberal values seem and the more open to difference liberal modernity declares itself, the more dismissive of difference it becomes and the more closed it seeks to make the circle of acceptability."[29] Accordingly, Goldberg traces the liberal impulse from Locke, Hume, Kant, and Mills, to contemporary theorists of rights, demonstrating where and how race is conceptually able to insinuate itself into the terms of each discursive shift.

Smith's argument assumes that American civic identity has always drawn from these three interrelated but analytically distinguishable ideologies. In this way Smith is able to address not only where these traditions mutually inform each other in the promotion of racist exclusions, but also where these traditions are in tension. For example, Smith acknowledges the ways in which a republican ideology, with its insistence on social homogeneity and small political communities, feeds racist exclusions. He also points to the crucial tension between liberalism's commitment to freewheeling individualism and the socially repressive elements of republican and ascriptive Americanist ideologies. Similarly, Smith highlights the tension between republicanism's emphasis on civic participation and duty to the polity and an ascriptive tradition that theorizes citizenship not in terms of one's capacities for "doing" but rather in terms of one's innate "being."

Of course, numerous scholars and critics have and will continue to protest that such ascriptive American impulses, like racisms in general, are more psychological than ideological—a mix of primal tribal loyalties with the fears and anxieties that accompany an encounter with the Other. What such theories of racism leave unexamined is the degree to which occurrences of racialized exclusion are in fact purposeful and quite rationally instituted for the aim of gaining political or economic power. Smith's insistence that ascriptive Americanism proved to be not only intellectually respectable but also a politically and legally authoritative discourse has been supported by historical evidence recently brought to light by John Higham, Reginald Horsman, George Fredrickson, David Theo Goldberg, Ivan Hannaford, Matthew Frye Jacobson, and others.[30]

As these scholars make clear, the discourse of race has been in circulation since the seventeenth century, but its ascendancy in the nineteenth century to a form of legitimate science posed dramatic challenges to the central tenets of classical republican and liberal political traditions, effecting changes not only in American political thought but European thought as well. In his impressive *Race: The History of an Idea in the West*, Ivan Hannaford explains how the centuries-long intellectual history of race—a discourse that increasingly came to identify itself with natural history, science, and thus modernity's interest in scientific forms of social amelioration or social engineering—challenged those notions of citizenship and political community, derived from antiquity, that laid the foundation for modern political thought. According to Hannaford,

the emergence of political life and law (*polis* and *nomos*) [in antiquity] was the outcome of a heated and controversial debate about words and letters (*logomachy*) in a public place (*agora*), which might lead to interesting solutions to the puzzles (*logogriph*) of human existence. One important suggestion arising from this discourse was that secular human beings might be persuaded to try a novel form of governance that provided options and alternatives to the prevailing forms of rule then surrounding them. *It was not a matter of Nature, but a difficult and original choice.*[31]

In contrast, from the end of the seventeenth century to the dawn of the twentieth, Hannaford observes that natural history increasingly became the basis for inquiry into legitimate forms of government—meaning that emphasis was placed on the temperament and character of races and the discovery of their true origins, rather than on political histories and the vices and virtues of actual states. Writers like Montesquieu, Hume, Blumenbach, Kant, Herder, and Burke contributed to the emergence of a self-conscious idea of race, and with the work of Niebuhr in the early nineteenth century "history was not the history of historical political communities of the Greco-Roman kind, but . . . transmogrifications of peoples into 'races' on a universal scale."[32] After the advent of Darwinism, Hannaford argues,

> it was generally agreed that classical political theory had little or nothing to offer Western industrial society. Notions of state drew support from the new literatures of nation and race. The tests of true belonging *were no longer decided on action as citizenship* but upon the purity of language, color, and shape. And since none of these tests could ever be fully satisfied, all that was left *in the place of political settlement* were ideas of assimilation, naturalization, evacuation, exclusion, expulsion, and finally liquidation.[33]

It is in this sense that Hannaford, after Michael Oakeshott, suggests that race must be understood as the perfect antonym for politics.

Similarly, capturing the rise of mid-nineteenth-century faith in race thinking and the simultaneous decline in the modern liberal commitments, Reginald Horsman observes that it had become "unusual by the late 1840s to profess a belief in innate human equality and to challenge the idea that a superior race was about to shape the fates of other races for the future of the world. To assert this meant challenging not only popular opinion, but also the opinion of most American intellectuals."[34] More recently, Matthew Frye Jacobson has addressed the ways in which racial science reformed common-sense understandings of the governing capacities of both nonwhite and white races. According to Jacobson, since the 1790 Naturalization Act, European immigrants had been granted entrance to the United States solely on the grounds of their whiteness. The unprecedented waves of immigrants in the mid-nineteenth century now caused concern that the policy was entirely too liberal and too inclusive. "Fitness for self-government," an attribute accorded exclusively to whiteness prior to the nineteenth century, now generated a "new perception of some Europeans' unfitness for self-government, now rendered racially in a series of subcategorical white groupings—Celt, Slav, Hebrew, Iberic, Mediterranean, and so on—white Others of a supreme Anglo-Saxondom."[35] Jacobson explains:

> It was the racial appellation "white persons" in the nation's naturalization law that allows the migrations from Europe in the first place; the problem this immigration posed to the polity was increasingly cast in terms of racial difference and assimilablity; the most significant revision of immigrant policy, the Johnson-Reed Act of 1924, was founded upon a racial logic borrowed from biology and eugenics.[36]

Thus, Jacobson further complicates the history of ascriptive Americanism as a civic ideology by reconceiving it as a response to the political crisis created by the over-inclusivity of the category "free white persons" in the 1790 naturalization law, and hence "a history of a fundamental revision of whiteness itself."[37]

In short, democratic, republican, and ascriptive ideologies, Smith argues, have always appeared on the historical stage in various combinations, as opposed to the dominance of any one in its "pure" or "ideal" form. And as the above allusion to the impact of racial science on late-nineteenth-century American political thought would indicate, Smith contends that various combinations of "liberal republicanism" dominated political agendas up to the

1870s; then a "republican nativist" agenda became more prominent. The hegemony of republican nativism only increased through the 1920s and persisted until the 1950s, when contemporary liberal ideas gained greater authority.[38]

We want to expand the implications of this important body of work on American civic identity by arguing that the reproduction of alternative conceptions of citizenship demands various forms of institutional support, particularly in educational apparatuses where the onus of responsibility for molding a competent and productive citizenry largely falls. If Smith is correct in his assessment of the general rearticulation of citizenship in the last decades of the nineteenth century, one would reasonably expect to see an equally profound curricular shift in higher education commensurate with the dramatic changes in the political thought that marked the era, especially given the university's historic role in the production of political and moral leadership. And in fact we do. Simultaneous with the transformation of the notion of citizenship and of political life more generally, American universities inaugurate a transition in the humanities from classical rhetoric to philology first, and then literary studies. The transition from rhetoric to English studies is significant, particularly when one considers how uniform and unchanging the college curriculum was until the late nineteenth century. Typically, undergraduate education centered on three to four years of required rhetoric courses in which students produced written essays and public addresses in the promotion of civic responsibility and political leadership. Yet, by the turn of the century the classical curriculum had all but disappeared, and English emerged as a new discipline. Overwhelmingly literary in orientation, the goals of the new curriculum were twofold: to produce an organic awareness of national cultural traditions that link Americanness with a specific version of whiteness, and to cultivate "discrimination," good taste, and moral sensibility—the latter objective as racially coded as the former.[39] As curricular emphasis shifted from the production of texts to their consumption, the arrival of literature as an object of formal study inaugurated not only the end of the classical university, but also a dramatic decline in public discourse and the practice of citizenship as an educational imperative.

The task at hand, then, is to demonstrate in a clear and concise way the differences between the classical curriculum and its modern counterpart in terms of how each negotiates the demands of a broader culture of politics—and participates in a politics of culture, particularly with respect to race. To do this, we will contrast briefly the educational thought of Thomas Jefferson and Calvin Coolidge, both of whom wrote on the civic function of higher education, and more particularly of the role of language and literature in the production of specific models of civic identity and national cultural tradition. We thus argue that the transition from rhetoric to literary studies is in part a function of changing definitions of citizenship, politics, race, and national identity, by contrasting Jefferson's plans for university education, as representative of the "liberal republican" ideological interests prior to the late nineteenth century, and Coolidge's program for higher education, as representative of the "nativist republican" agenda that marked the era from the 1870s to the 1950s. In spite of our characterization of the rise of literary education as a "fall" from public grace, our point is not to argue for a simple "return" to rhetoric, but to demonstrate how forms of race consciousness informed both the classical curriculum and its literary counterpart.

Not only is race a central determination in the history of liberal arts education; it is also central to its future. Race cannot be addressed as a discourse removed from mainstream educational theory, a burden imposed from the outside by the forces of multiculturalism or "PC." First, in order to grasp the significance of the rise of literary studies in the liberal arts curriculum, it is important to understand the rhetorical tradition that was in place before its decline.

From Rhetorical to Literary Education

Progressive scholars such as Raymond Williams, Terry Eagleton, James Berlin, and Sharon Crowley have tended to explain the simultaneous decline of classical rhetoric and the rise of literary education in terms of emergent bourgeois class interests.[40] Williams argues, for example, that the turn to literature is best understood as a major affirmative response, in the name of human creativity and imagination, to the socially repressive and mechanistic nature of the new capitalist order. To understand this profound curricular shift in the nineteenth-century college, scholars must address the advance of capitalism *and* the impact of mass immigration, the influence of the leading scientific discourses of the day such as social Darwinism, efficiency, eugenics, and the nation's commitment to racial segregation. These events, as we've already indicated, induced dramatic reconceptualizations of liberal political philosophy, national identity, citizenship, and race—all of which affected educational thought and practice. In other words, to the degree that the political order was rearticulated in terms of a natural order, citizenship became less contingent on one's performance in public life and more on an innate capacity determined by blood and heredity. This is not to suggest that civic education was altogether abandoned; rather we argue that the kind of citizen university curricula attempted to put into place was radically reconfigured. If the goal of classical rhetorical education was to enhance the practice of citizenship as a performance of duties and responsibilities to the political community in exchange for rights and entitlements in keeping with liberal and republican ideologies, the new educational mandate privileging literary study was, at least in part, an attempt to establish an ascriptive notion of citizenship—by redefining it not as a function of "doing," but one of "being." Thus it became the "duty" of students endowed with the appropriate class and racial inheritance simply to receive, appreciate, and protect their distinctive ethnocultural heritage.[41]

According to historians of rhetorical and literary education such as James Berlin, S. Michael Halloran, and Gerald Graff, rhetoric was at the center of a relatively stable and unchanging college curriculum prior to the late nineteenth century. Since their appearance in the seventeenth and eighteenth century, American colleges followed the traditions established by Oxford, Cambridge, and the continental universities in the preparation of their overwhelmingly white male student body for law, ministry, medicine, and politics.[42] Rhetoric was emphasized so heavily in these disciplines because, as Halloran explains,

it was understood as the art through which all other arts could become effective. The more specialized studies in philosophy and natural science and the classical languages and literatures would be brought to focus by the art of rhetoric and made to shed light on problems in the world of social and political affairs. The purpose of education was to prepare men for positions of leadership in the community, as it had been for Cicero and Quintilian.[43]

Investigating more specifically the various ways in which rhetorical education was conceived in the classical college, Halloran has argued that in contrast to the anticlassical bias in the seventeenth-century college, classical rhetoric as the art of public discourse flourished in the eighteenth century at Harvard and newer colleges such as William and Mary and Yale. For Halloran, the tradition of classical rhetoric gave "primary emphasis to communication on public problems, problems that arise from our life in political communities."[44] The emergence of the classical impulse was reflected in the increasing curricular emphasis on the English language and on effective oral communication, which dealt with public issues and concerns, a shift Halloran attributes to the greater availability of works by Cicero and Quintilian during the second decade of the eighteenth century.

A graduate of William and Mary in the mid-1760s, Thomas Jefferson wrote extensively on the relationship between higher education and the political life of the nation, and his views are clearly reflective of the classical training he received there. In fact, Jefferson's vast educational plans for a free and universal, multi-tiered educational system including primary, grammar, and university training are central to his social and political thought. For Jefferson, education was the primary means for producing the kind of critically informed and active citizenry necessary to both nurture and sustain a democratic nation; he argued, in keeping with classical republican tradition, that democracy was the highest form of political organization for any nation because it provided the conditions for its citizens to grow both intellectually and morally through the exercise of these faculties. In addition to three legislative proposals—the Bill for the More General Diffusion of Knowledge; the Bill for Amending the Constitution of the College of William and Mary, and Substituting More Certain Revenues for Its Support; and the Bill for Establishing a Public Library—which constitute the core of his educational thought, Jefferson elaborated his educational vision in his *Notes on the State of Virginia* and in numerous private letters to his nephew Peter Carr and others. Jefferson's classic preamble to the 1776 Bill for the More General Diffusion of Knowledge bears the hallmark of his views on the relationship between education and public life:

> Whereas . . . certain forms of government are better calculated than others to protect individuals in the free exercise of their natural rights . . . experience hath shewn, that even under the best forms, those entrusted with power have . . . perverted it into tyranny; and it is believed that *the most effectual means of preventing this would be, to illuminate, as far as practicable, the minds of the people at large*; And whereas it is generally true that people will be happiest whose laws are best, and are best administered, and that laws will be wisely formed, and honestly administered, in proportion as those who form and administer them are wise and honest.[45]

As this passage indicates, "illuminating" via formal education is central to Jefferson's liberal philosophical leanings and his republican agendas; it is both a means for preserving individual rights and property from all forms of tyranny and a means for enabling wise and honest self-government. What both traditions share, as is evident in Jefferson's prose, is a conception of education as a preeminently political issue—and politics as a preeminently educational issue. (As we will demonstrate shortly, Jefferson's thought was also reflective of his ascriptive agendas, as his role in the nation's legacy of racialized exclusion makes clear.) After his administration, he penned a Bill for Establishing a System of Public Education and the Report of the Commission Appointed to Fix the Site of the University of Virginia, commonly known as the Rockfish Gap Report. In this 1818 document, Jefferson maps the objectives for university education and provides an eloquent defense of higher education as a public good worthy of federal funding. According to Jefferson, the purpose of higher education is to provide the following (which we quote at length, if only to underscore how little in common the contemporary mission of the corporate university has with its historic counterpart):

> To form the statesmen, legislators and judges, on whom public prosperity and individual happiness are so much to depend;
>
> To expound the principles and structures of government, the laws which regulate the intercourse of nations, those formed municipally for our own government, and a sound spirit of legislation, which banishing all arbitrary and unnecessary restraint on individual action, shall leave us free to do whatever does not violate the equal rights of another;
>
> To harmonize and promote the interests of agriculture, manufactures and commerce, and by well informed views of political economy to give a free scope to the public industry;
>
> To develop the reasoning faculties of our youth, enlarge their minds, cultivate their morals, and instill into them the precepts of virtue and order;
>
> To enlighten them with mathematical and physical sciences, which advance the arts, and administer to the health, the subsistence, and the comforts of human life;
>
> And, generally, to form them to habits of reflection and correct action rendering them examples of virtue to others, and of happiness within themselves.[46]

As these objectives indicate, the branches of higher education are responsible for producing effective moral and political leadership, not trained technicians; where professional interests are alluded to, they are always tied to the interests and well-being of the commonweal. In contrast to the current state of affairs, there is no confusion between education and training.[47]

Jefferson divided the university curriculum into ten branches: ancient languages, modern languages, five branches of mathematics and the sciences, government, law, and finally "ideology," which included studies in grammar, ethics, rhetoric, and belles lettres. Private letters to his nephew and protégé, Peter Carr, indicated more specifically what the study of ideology entailed. He advised Carr to read ancient history including works by Herodotus, Thucydides, Xenophontis Hellenica, Xenophontis Anabasis,

Quintus Curtius, and Justin; Roman history, then modern; Greek and Latin poetry including Virgil, Terence, Horace, Anacreon, Theocritus, and Homer; and moral philosophy. According to Jefferson, such readings provide ordinary citizens "knowledge of those facts, which history exhibiteth, that possessed thereby of the experience of other ages and countries, they may be enabled to know ambition under all its shapes, and prompt to exert their natural powers to defeat its purposes."[48] The pedagogical emphasis here is on the production of an active and critical citizenry skilled not only in the protection of their individual rights but popular participation in the interests of self-governance. "If the condition of man is to be progressively ameliorated," Jefferson argued, "education is to be the chief instrument in effecting it."[49] It is interesting to note, in light of the direction that rhetorical education would take, that Jefferson also advises his protégé to read Milton's *Paradise Lost*, Ossian, Pope, and Swift "in order to form your style in your own language."[50] These literary works were recommended as models for the improvement of form in oral and written communication—and not, as they would later be proffered, for honoring one's racial heritage or asserting racial superiority.

Thus Jefferson inevitably looked to education as a means of social, moral, and political uplift, as well as an aid to the personal and professional advancement of individual citizens. He hoped that formal educational experience would lead, by force of habit, to learning as a lifelong practice. "Education generates," he insisted, "habits of application, of order, and the love of virtue; and controls, by the force of habit, any innate obliquities in our moral organization."[51] In other words, education secured the progress of "man":

We should be far, too, from the discouraging persuasion that man is fixed, by the law of this nature, at any given point; that his improvement is a chimera. . . . As well might it be urged that the wild and uncultivated tree, hitherto yielding sour and bitter fruit only, can never be made to yield better; yet we know that the grafting art implants a new tree on the savage stock, producing what is most estimable in kind and degree. Education, in like manner, engrafts new man on the native stock, and improves what in his nature was vicious and perverse into qualities of virtue and social worth.[52]

But Jefferson was not interested in the rights, civic participation, or general progress of women in general and men of color. Jefferson's views on both women and African Americans are now well known. The statesman who penned the Declaration of Independence and proclaimed universal human rights and human equality also insisted that, unlike Native Americans, African Americans did not have the natural intellectual endowment necessary for self-government.[53] In his *Notes on the State of Virginia*, Jefferson wrote that "Comparing them [African Americans] by their faculties of memory, reason, and imagination, it appears to me that in memory they are equal to whites: in reason much inferior, as I think one could scarcely be found capable of tracing and comprehending the investigations of Euclid; and that in imagination they are dull, tasteless, and anomalous."[54] As if in anticipation of the eugenic vision of Coolidge a century later, Jefferson also argued that "amalgamation with the other color produces a degradation to which no lover of his country, no lover of excellence in the human character can innocently consent."[55] All his major proposals for free public education excluded slaves. And, as in classical Greece, Jefferson held that women belonged in the private or domestic sphere and not in public life; as citizenship was a male privilege, females were provided schooling only at the elementary level.

As these exclusions make clear, Aristotle was correct in assuming that a good citizen is not the same as a good man; in fulfilling the demands of their polity, citizens are only as good as the laws that they frame and obey. Any attempt to reappropriate elements of a "classical" rhetorical education, with its emphasis on the responsibilities of citizenship and the importance of participation in public life, will have to engage the ways that citizenship and agency itself—defined in terms of fitness for self-government—have been both gendered as male and racially coded as white since the nation's inception.

Jefferson's 1818 commentary on higher education is in keeping with classical liberalism's faith in natural law, rationality, freedom, and the ameliorative force of social institutions such as education. Within the next hundred years these "classical" liberal tenets underwent a profound revision in response to rapidly changing social and political conditions, as well as the Darwinian revolution in scientific thought. Unlike Jefferson's faith in the average citizen's capacity to reason, debate, and take action in the interests of justice and the public good, the "modern" search for truth required scientific method and the intervention of expert knowledge. Jefferson's beliefs that human reason would triumph over the base instincts of human nature and that social progress was inevitable were significantly challenged by modern scientific findings. Influenced in part by Charles Darwin's observations that some species decline while the fit survive, and in part by the crises brought about by rapid urbanization and industrialization—overcrowding, poverty, disease, crime, revolt—modern liberals no longer believed that progress was inevitable, but that it required expert social planning and scientific management. Moreover, in contrast to Jefferson's commitment to intellectual and moral growth through education, modern thought held that such improvement was limited by genetic endowment.

Lawrence Cremin explains the influence of Charles Darwin and Herbert Spencer on educational thought and practice in the following terms:

> because the development of the mind followed evolutionary processes and because evolutionary processes worked themselves out over time, independent of immediate human acts, education could never be a significant factor in social progress. The only thing teachers could do was provide the knowledge that would enable people to adapt to their circumstances.[56]

Specifically, Spencer's *Education: Intellectual, Moral and Physical* was used to legitimate the transition from the classical curriculum to a version of "progressive" education associated with the work of Harvard president Charles A. Eliot and the National Education Association's Committee of Ten. It was in part through Eliot's efforts that the classical curriculum was eventually replaced by a differentiated course of study designed to help the nation's youth adapt to their environment rather than shape or reform it.[57] Alarmed at the increasing ethnic diversity of the school environment and convinced of the intellectual incapacities of all but "pure American stock" (which excluded all those white races that came to the

United States in the second wave of European immigration), Eliot became a staunch advocate for vocational education.

In 1908 he suggested that modern American society was made up of four largely unchanging social classes: a small leading class, a commercial class devoted to business interests, skilled craftsman, and "rough workers." Failure to recognize these divisions, according to Eliot, resulted in an inefficient system in which "the immense majority of our children do not receive from our school system an education which trains them for the vocation in which they are clearly destined."[58] Once an advocate of liberal education for all youth, Eliot pushed for a differentiated curriculum appropriate to the largely "innate" capacities of various classes and races. In the same year, fellow Spencerian Alfred Schultz captured the race consciousness so influential in educational reform as he bemoaned the limits of assimilation in the following analogy:

> The opinion is advanced that the public schools change the children of all races into Americans. Put a Scandinavian, a German, and a Maygar boy in at one end, and they will come out Americans at the other end. Which is like saying, let a pointer, a setter, and a pug enter one end of a tunnel and they will come out three greyhounds at the other end.[59]

What Schultz's startling pronouncement reflects is an increasingly mainstream concern over the impossibility of Americanization for some (in this instance, white) immigrant races. In fact, some races were agents of de-Americanization, meaning that their presence threatened the purity of the gene pool of "real American stock." To understand how pervasive such race thinking was in the first decades of the twentieth century, it is interesting to note the similarity in thought between intellectuals like Eliot and Schultz and Klansmen like Imperial Wizard Hiram Wesley Evans, who over a decade later insisted that federal legislation must be passed to keep out delinquent and downtrodden races from the Mediterranean and Alpine regions in these terms: "We demand a return of power into the hands of the everyday, not highly cultured, not overly intellectualized but entirely unspoiled and not de-Americanized average citizens of the old stock."[60]

Indeed, evidence of such race thinking would shortly find its way to the executive branches of government. In a 1921 article in *Good Housekeeping* entitled "Whose Country Is This?" Vice President Calvin Coolidge put the weight of his support behind ascriptive Americanist legislative agendas, rationalizing his endorsement by invoking the same rhetoric as Jefferson—the goal of inculcating good citizens. Now the production of good citizens was less a matter of civic education than one of social engineering— an attempt to govern through the logic of scientific management and efficiency. And what better place to issue advice in the national interest to American mommies and daddies than *Good House-keeping?* In short, Coolidge's plan meant subjecting citizens to a process of Americanization, which was only possible with those groups, or "races," of people capable of self-government (and thus full assimilation) in the first place. Thus, with the racial science of the day behind him, Coolidge declared that "Biological laws tell us that certain divergent people do not mix or blend. The Nordics propagate themselves successfully. With other races, the outcome shows deterioration on both sides."[61] He concluded in favor of leg-islation restricting the flow of immigrants of non-Nordic origins, stating, "Quality of mind and body suggests that observance of ethnic law is as great a necessity to a nation as immigration law."[62] The ascriptive Americanist agendas to which Coolidge subscribes reduces the complexities of citizenship to the question of member-ship, which is determined on the basis of heredity and ignores altogether issues of citizens' rights, civic duties, and political par-ticipation in the community.

As we've already indicated, such a limited notion of citizenship is in part the result of the declining faith in civic institutions as a whole, which accompanied the growing influence of racial science. Though the origins of race thinking hardly begin with Darwin and Spencer, their work spawned an intellectual movement in which human society and politics were understood to be subject to the same rules of evolution that applied to the natural world. Thus, as Hannaford has argued, it provided a scientific rationale for decry-ing Aristotelian political theory and all aspects of the Greco-Roman polity as out of step with modernity. Society was now understood to be "a natural entity in a state of war in the classic

Hobbesian sense, in which power and force in the hands of the classes or races, scientifically applied, would lead inevitably to the progressive ends of... 'industrial civilization.'"[63] Accordingly, by the mid-1850s notions of legal right, treaty, compromise, settlement, arbitration, and justice, which constitute political community, were "eclipsed, and then obliterated" by a doctrine of "natural evolutionary course" that expressed itself in a language of "biological necessity, managerial efficiency, and effectiveness."[64]

Coolidge did, however, argue for the necessity of higher education, though in vastly different terms than Jefferson. According to Coolidge, the "first great duty" of education was "the formation of character, which is the result of heredity and training."[65] Whereas Jefferson's educational thought bore the legacy of Enlightenment racism, Coolidge's flirted with eugenics. While the passing of the Johnson Act was a great victory for Coolidge's administration, he told the National Education Association that such legislation was, in the final analysis, of secondary importance. National progress depended not on the "interposition of the government" but on "the genius of the people themselves."[66] Real appreciation of this "genius" required more "intense" study of our "heritage," and particularly "those events which brought about the settlement of our own land."[67]

Curiously, Coolidge's referent was not the Revolutionary Era and the end of English colonial domination. "Modern civilization dates from Greece and Rome," he argued, and just as they were "the inheritors of a civilization which had gone before," we were now their inheritors.[68] In answering the question, "What are the fundamental things that young Americans should be taught?" Coolidge responded, "Greek and Latin literature."[69] Coolidge's response gives rise to two apparent contradictions: first, real "American stock" was not of Greek origin—though he locates the origins of national culture there—and, as if to keep it that way, the Johnson Act restricted the real descendants of classical Greece from U.S. citizenship. The latter contradiction is easily resolved. According to the *Dictionary of Races and Peoples*, which comprised volume nine of the Dillingham Commission's Report on Immigration and was presented to the sixty-first Congress in December 1910, modern Greeks were themselves a different race from the ancients, and now a "degenerate" population as a result of the Turkish invasion and subsequent amalgamation. Hence, the former contraction, insisting on our Greek origins, unfolds: Americans were the inheritors of civilization not because "we" descended racially from the ancient Greeks, but because we remained, as the Johnson Act would ensure, a pure race. Thus one witnesses in Coolidge's social and educational policy the same fear of racial amalgamation to which Jefferson gave voice. "Culture is the product of a continuing effort," Coolidge asserted, because "The education of the race is never accomplished."[70] The process of educating the nation's citizenry to understand and take pride in their racial and cultural inheritance was ongoing because its purity was continually threatened by unassimilable races. In short, Coolidge's support for the study of Greek and Roman literature is for vastly different reasons than Jefferson's. For Jefferson it was about learning how to take an active and ongoing role in democratic public life; for Coolidge, it was about the appreciation and protection of one's racial endowment through the harnessing (or educating) of desire in the name of individual morality and patriotism.

David Shumway has situated the shift from rhetoric to literature in the period when "historians first produced the Teutonic-origins theory of American civilization, that Anglo-Saxonism and Anglophilia reached its peak among the American cultural elite and that concerted efforts were made to Americanize immigrants."[71] In such a climate, the turn to literature was quite natural. As Shumway explains, "Literature was more than peripherally related to this racism since it was widely held that literature expresses the essential character of a race. This is true because language, the substance of literature, 'is an expression or function of race.'"[72] But as we have attempted to show, "Anglo-Saxonism" did more than influence literary conversations; it also changed the ways in which broader concepts such as the nation, politics, civic duty, citizenship, and civic education were understood. Indeed, the forms of race thinking that gave rise to racist exclusions have flourished throughout the entire modern period, and continue to exert their influence today. Covering centuries rather than decades, the influence of racist thought and practice on civil institutions cannot be reduced to the "Anglo-Saxon mystic" or "Anglophilia" of the turn of the century, as if such institutions were now untouched by the politics of race.

What the comparison between Jefferson's and Coolidge's educational thought suggests is that different versions of citizenship—liberal democratic, civic republican, and ascriptive Americanist—presuppose a curricular and pedagogical model that puts into place subjectivities invested with specific notions of identity and community, knowledge and authority, values and social relations. Additionally, each pedagogical model makes claims on particular forms of consciousness, memory, and agency that influence not only individual subjects but the collectivity as a whole.

It is possible to assess critically each model as it circulates in contemporary conversations about the future direction of liberal arts education, analyzing how the relationship between pedagogy and politics is both theorized and enacted by posing the following questions. First, what are the conditions for the development of both individual and collective agency? Or, put in slightly different terms, how is learning linked to civic action or social change? Do citizens learn to take an active role in self-government, or is the educational agenda one of adaptation or subordination? Second, how is knowledge produced? Is it dialogical and open to critique or is it canonical and sacred, and so above criticism? Who controls the production of knowledge and who benefits from it? Third, how does each model of pedagogy legitimate different versions of social relations—democratic relations and hierarchical ones? Is the notion of political community that such curricular models and pedagogies give rise to marked by inclusion or exclusion? Fourth, does the pedagogical model make clear the grounds for its own authority or is it considered natural, innate, or prepolitical? Finally, what values are legitimated? Are social homogeneity and consensus privileged? Or are difference and dissent? Obedience, or the questioning of authority? With these issues in mind, and in light of the ways in which different versions of citizenship have been articulated to educational policy, we would like to turn to the contemporary debates over civic education.[73]

The Contemporary Contest over Civic Education

The narrative we've provided of the past and future of English studies is not what Bruce Robbins would rightly dismiss as a "narrative of the fall," of the discipline's (even the humanities') retirement from public life. Rather, our ongoing interest has been in demonstrating how changes in notions of liberal democratic politics, nationalism, citizenship, and always closely associated notions of race bring about corresponding shifts in educational thought and practice. The early transition from classical rhetoric to literary study, which shifted emphasis from civic to aesthetic concerns, is really about trading one form of citizenship for another— one participatory and public, the other nationalistic and privatized.

As the intellectual basis for theorizing civic capacities in terms of race faltered and gave way beginning in the 1930s, the discourse of literature, now grounded in theories of cultural value as opposed to racial heritage, was refashioned into a highly formalized, insular, and "professional" rhetoric that prided itself on its distance from public life. The "New Criticism," whose heyday spanned from the 1930s to the 1950s, no longer derived its authority from a direct socially ameliorative function, but rather from its withdrawal into disinterestedness. Of course, there remained a vast distance between the rhetoric of disinterestedness and professional neutrality and its actual practice in both scholarship and pedagogy. The social lessons of race-based literary study could be achieved, perhaps even more efficiently, with a rhetoric of pristine objectivity rather than social engineering. The privileged question of "value" within the New Critical lexicon still wielded its profoundly Eurocentric and exclusionary force, and the supreme legitimating discourse remained the rhetoric of science, purged of its former commitments to racial theory.

With the upheavals of the 1960s, needless to say, the era of high modernist hegemony came to a close and literature and its critics once again had to renegotiate their relationship to public life. To say the least, the task has been far from easy. While we won't recount here the myriad criticisms of the postmodern, multicultural academy, the upshot of these debates is to have produced a new genre of writing exclusively concerned with the future direction of English studies in particular, and the humanities more generally. Some of these debates have mourned the alleged passing of literature altogether. Former provost at Yale and dean of the graduate school at Princeton Alvin Kernan commenced the eulogistic theme popular among conservatives with *The Death of Literature* (1990). This was quickly followed by melancholy tomes like Harold Bloom's *The Western Canon*, with its opening "Eulogy" (1993); John Ellis's *Literature Lost: Social Agendas and the Corruption of the Humanities* (1997); Kernan's edited collection, *What's Happened to the Humanities?* (1997); and Roger Shattuck's *Candor and Perversion: Literature, Education and the Arts* (1999). Although the mood is unmistakably *in memoriam*, the arguments attempt to establish the rationale for a return to the good old days of aesthetic formalism and closed canonicity.

As space will not allow a full investigation of the claims made in the above literature, the general sentiment and mode of critique of this subgenre can best be captured in a pithy commentary in a September 1996 issue of the *National Review*. Senior editor and Dartmouth professor Jeffrey Hart announced that something was terribly amiss in higher education and had been for at least a decade. He likened the discovery to an occasion in W. H. Auden where a guest at a garden party senses disaster and discovers a corpse on the tennis court. What has so profoundly disturbed the country-club serenity of the Ivy League? To begin with, Hart attests, recent intellectual trends such as postmodernism and multiculturalism, as well as their corollary in policy, affirmative action. "Concomitantly," he adds, "ideology has been imposed on the curriculum to a startling degree."[74] Nonetheless, Hart assures his readers that all is not lost. And as the title of the essay, "How to Get a College Education," forecasts, he offers the following advice to undergraduates:

> Select the ordinary courses. I use ordinary here in a paradoxical and challenging way. An ordinary course is one that has always been taken and obviously should be taken—even if the student is not yet equipped with a sophisticated rationale for so doing. The student should be discouraged from putting his money on the cutting edge of interdisciplinary cross-textuality.
>
> Thus, do take American and European history, an introduction to philosophy, American and European literature, the Old and New Testaments, and at least one modern language. It would be absurd not to take a course in Shakespeare, the best poet in our language. . . .

I hasten to add that I applaud the student who devotes his life to the history of China or Islam, but that...should come later. America is part of the narrative of European history.

If the student should seek out those "ordinary" courses, then it follows that he should avoid the flashy come-ons. Avoid things like Nicaraguan Lesbian Poets. Yes, and anything listed under "Studies," any course whose description uses the words "interdisciplinary," "hegemonic," "phallocratic," or "empowerment," anything that mentions "keeping a diary," any course with a title like "Adventures in Film."

Also, any male professor who comes to class without a jacket and tie should be regarded with extreme prejudice unless he has won a Nobel Prize.[75]

At first glance, it is easy to disregard Hart's polemical essay as so much right-wing hysteria. But the challenges posed to these academic "fads" are hardly confined to conservative circles alone and so cannot be dismissed as *merely* ideological. In the 1990s, for example, a number of progressives have denounced the cultural left, as Ellen Willis points out, for "its divisive obsession with race and sex, its arcane 'elitist' battles over curriculum, its penchant for pointy-headed social theory and its aversion to the socially and sexually conservative values most Americans uphold."[76] In his *Professional Correctness*, Stanley Fish takes to task the literary critic who would conclude his analysis of *Sister Carrie* or the *Grapes of Wrath* with a commentary on homelessness rather than with an assessment of literary realism and assume it will find its way to the Department of Housing and Urban Development. Exposing as fallacious and insipid any academic pretense to social change, Fish advocates a return to the practical and professional criticism associated with John Crowe Ransom and the New Critics of the 1940s. In short, he argues that the contemporary push for English studies to become cultural studies threatens the integrity of the "kind of thing we [allegedly] do here," which, according to Fish, is about the aesthetic reading of canonical texts, a judgment with which Hart would agree.[77] Further, the loss of "distinctiveness" of what "we do" in English threatens to undermine the discipline's *raison d'être*.

Similarly, in "The Inspirational Value of Great Works of Literature," Richard Rorty suggests that the current academic fervor for literary analysis of the "knowing, debunking, *nil admirandi* kind" drains the possibilities for enthusiasm, imagination, and hope from scholars and students alike. In place of critical analysis, Rorty urges an appreciation of "great" works of literature; by that he means seeking inspiration from works of literature that "inculcate the same eternal 'humanistic' values."[78] What such an appeal to transcendent truth means coming from a philosopher committed to the notion of cultural relativism remains unclear, but the universalizing gesture has a profoundly Eurocentric pedigree. Decrying the rise of cultural studies in English departments and its cult of knowingness, Rorty contends that "You cannot ... find inspirational value in a text at the same time you are viewing it as a product of a mechanism of cultural production."[79] Pitting understanding against the romantic values of awe, inspiration, and hope, Rorty advocates a kind of intellectual passivity among readers in the name of hopefulness. Basically it is helplessness.

It is worth noting that Lynn Hunt made a similar claim that "cultural studies ... may end up providing the deans with a convenient method for amalgamating humanities departments under one roof and reducing their faculty size."[80] According to this logic, it is theoretical discourses associated with cultural studies rather than the logic of corporatization and downsizing that challenges the continued existence of the humanities. What both Rorty and Fish share with conservatives such as Hart, Bloom, Hunt, and others is a desire to narrow the field of intellectual inquiry, to reduce literary interest to what makes it most "distinctive": its capacity for formal aesthetic appreciation. Such a call is a retreat from the political in the name of professional survival. The moral and ethical imperative to engage the social implications of how students learn to read is traded for either a breathless romanticism (Rorty) or a cool-headed pragmatism (Fish). Edward Said is not the first intellectual to associate the call for professionalism and its attendant demands for specialization and expertise with intellectual inertia and laziness. In the study of literature, Said argues, "specialization has meant an increasing technical formalism, and less and less of a historical sense of what real experiences actually went into the making of a work of literature."[81] The result is an inability to "view knowledge and art as choices and decisions, commitments and alignments, but only in terms of impersonal theories or methodologies."[82]

Moreover, criticisms by Hart, Fish, Rorty and others resonate powerfully with the growing concerns of many undergraduate populations over politically correct curricula, diversity requirements, and teachers who assume that race, class, and gender are the only analytical tools for engaging cultural texts. These are the very students who are supposed to feel more empowered, critically literate, and socially conscious through their encounter with these discourses. So for the latter reason alone, it is necessary to engage Hart's depiction of the contemporary "multicultural turn" in university education as a kind of representative critique and offer a response.

While there is much to oppose in Hart's essay, some of his basic assumptions and concerns hold merit and warrant further analysis. First, Hart's repeated rant against courses like "Nicaraguan Lesbian Poets" and identity politics in general is one that—for vastly different reasons—gives intellectuals across the ideological spectrum some pause. While for conservatives such as Hart curricula gave way to the horror show of "political correctness" across university campuses in the 1980s and 1990s, progressives have critiqued its tendency to reproduce facile, oftentimes reactionary, understandings of the complexities of identity and the politics of race and gender—hence Keith Gilyard's recent insistence that the necessity for *theorizing race* now be taken seriously in rhetoric and composition. Such practices not only undermine complex notions of identity as multiple, shifting, and in process; they parade under the banner of a form of multicultural education that Stuart Hall criticizes for reproducing "essentialized notion[s] of ethnicity," gender, and sexuality.[83]

Second, the vast majority of scholars—even those in cultural studies, postcolonial studies, and women's studies—share Hart's commitment to providing students with an introduction to the intellectual traditions that have shaped contemporary culture. But unlike Hart, such scholars approach the question of content dialogically. That means, according to Stanley Aronowitz, they distinguish between the hegemonic culture, which constitutes the conventional values and beliefs of society, and subordinate cultures, "which often violate aspects of this common sense." Nor do they "assume the superiority of the conventional over the alternative or oppositional canon, only its power"; in short, they substitute the practice of critique for reverence.[84] Homi Bhabha has described the necessity for educators to promote critical literacies by teaching students to

> intervene in the continuity and consensus of common sense and also to interrupt the dominant and dominating strategies of generalization within a cultural or communicative or interpretive community precisely where that community wants to say in a very settled and stentorian way: this is the general and this is the case; this is the principle and this is its empirical application as a form of proof and justification.[85]

In contrast to Hart's emphasis on the transmission of "depoliticized" content, which rejects the need for educators to make explicit the moral and political thrust of their practices, real higher learning for Aronowitz and Bhabha takes up the task of showing how knowledge, values, desire, and social relations are always implicated in power. What these theorists share is an awareness that knowledge is not only linked to the power of self-definition, but also to broader social questions about ethics and democracy. Similarly, Paulo Freire insightfully argues that the "permanent struggle" that educators must wage against forms of bigotry and domination does not take the place of their responsibilities as intellectuals. He concludes,

> I cannot be a teacher without considering myself prepared to teach well and correctly the contents of my discipline, I cannot reduce my teaching practice to the mere transition of these contents. It is my ethical posture in the course of teaching these contents that will make the difference.[86]

Thus, in spite of Hart's compulsive use of the term, there is nothing "ordinary," historically given, or apolitical about the course of study he and a score of others from Harold Bloom to Richard Rorty propose for undergraduate education.[87] In fact, Hart's overzealousness betrays his efforts to legitimate such selections through an appeal to a version of common sense that is increasingly open to question; his obsessive iteration of "ordinary" reveals that such assumptions can hardly be taken for granted. Quite to the contrary, the selection of courses and topics Hart mentions have not "always" been taken; some, in fact, have been added to university curricula relatively recently. The study of Shakespeare, for example, is only as old as the English department itself, which has been around for slightly over one hundred years, when it displaced a much older tradition of classical rhetoric.

Finally, Hart's assessment of the essential function of a liberal arts education is a judgment with which few scholars could disagree. "The goal of education," he asserts, "is to produce the citizen."[88] At first glance, Hart's insistence that citizenship is the goal of higher education seems paradoxical, particularly in light of his pronouncement that ideology has denigrated academic pursuits. How is it possible, after all, to decouple civic education from the broader culture of politics? The answer to this apparent irony lies in Hart's definition of "the citizen," which abstracts civic membership from active, public performance in the interests of the commonweal. According to Hart,

> the citizen should know the great themes of his civilization, its important areas of thought, its philosophical and religious controversies, the outline of its history and major works. The citizen need not know quantum physics, but he should know that it is there and what it means. Once the citizen knows the shape, the narrative, of his civilization, he is able to locate new things—and other civilizations—in relation to it.[89]

Hart's citizen is a passive bearer of national cultural traditions, here made identical to those of Western culture. It is a far cry from the Aristotelian model of the virtuous citizen who "lives in and for the forum," actively pursuing the public good with single-minded devotion—a model that has always haunted republican notions of American civic identity.[90] She does not even have to master this knowledge, but rather, in game show–like fashion, be able to name it and know it's there. Republican emphasis on constant and direct involvement in governing as well as being governed, on duties and responsibilities in reciprocity, remain untheorized and, one assumes, unimportant to Hart's civic and educational vision.

Similarly, Hart's definition of citizenship is at odds with the liberal version of American civic identity. In contrast to most other nations for whom "national identity is the product of a long process of historical evolution involving common ancestors, common experiences, common ethnic background, common language, common culture, and usually common religion," American civic identity has historically been based on "political ideas," on an allegiance to the "American Creed" of liberal democracy.[91] Yet Hart's definition of the citizen is precisely based on the "common ancestors, common experiences, common ethnic background, common language, common culture, and usually common religion" that Huntington ascribes to other renditions of national identity. It is thus a direct descendant of the ascriptive Americanism dominant at the turn of the twentieth century. As such, it is a form of citizenship that offers no theory of politics because it cannot deal with notions of conflict or antagonism. Insisting on a common culture that promotes harmony on the basis of social homogeneity, it requires the exclusion of dissent and difference.

In spite of its deviation from common republican and liberal conceptions of citizenship, the definition Hart relies upon has nonetheless been a popular one in the contemporary debate over liberal arts education. For example, the notion of citizen as bearer of common cultural knowledge has been powerfully articulated by such scholars as E. D. Hirsch and Roger Shattuck. In his now classic 1987 volume, *Cultural Literacy: What Every American Needs to Know*, Hirsch argues that:

> Literate culture has become the common currency for social and economic exchange in our democracy, and the only available ticket to full citizenship. Getting one's membership card is not tied to class or race. Membership is automatic if one learns the background information and . . . linguistic conventions.[92]

The language Hirsch uses to describe national civic identity bears a striking resemblance to Hart's. Both scholars rely heavily on the criteria of common knowledge (and hence, common culture and experience) for civic membership, while at the same time claiming that conditions of inheritance—such as one's gender, race, or socioeconomic status, for in many ways the latter is inherited in spite of the myth of class mobility—are not prerequisites. Yet the knowledge Hart and Hirsch require of citizens is, nonetheless, race- and class-specific.[93] Like the nativist arguments at the turn of the century, their understanding of national cultural identity not only privileges a Eurocentric perspective of history and culture but also silently equates Americanness with whiteness in the interests of promoting an allegedly time-tested, Western "Great Books" curriculum that in actuality has only been around for little more than 80 years.[94] The similarity between the language conservatives like Harold Bloom uses to defend American cultural traditions with the eugenicist language of Calvin Coolidge is unmistakable:

> We [the United States of America] are the final inheritors of Western tradition. Education founded upon the *Iliad*, the Bible, Plato and Shakespeare remains, in some strained form, our ideal, though the relevance of these cultural monuments to life in our inner cities is inevitably rather remote.[95]

Like that of Coolidge, Bloom's rhetoric not only summons up a genealogy that links ancient Greece to modern American culture, but also establishes the vast distance between the final inheritors of Western European cultural traditions and the "inevitable" remoteness of our inner cities as a racial, as opposed to spacial, divide.

More recently Roger Shattuck, former president of the Association of Literary Scholars and Critics, lambasted educators and school boards alike for attempts to foster critical thinking over instilling well-defined content requirements reflective of a "core tradition" in the humanities. In English, the arts, and foreign languages, Shattuck claims, "the emphasis falls entirely on what I call 'empty skills'—to read, to write, to analyze, to describe, to evaluate."[96] How Shattuck proposes students engage a "core tradition" without recourse to such "empty skills" remains unclear—unless, like Hart, he feels that students "need not know" what (or how) texts like *Moby Dick* mean, only that they are simply "there."

Not only do the advocates of an Anglo common culture rely on transmission theories of pedagogy; they advocate, as Shattuck asserts, that "our schools will serve us best as a means of passing on an integrated culture, not as a means of trying to divide that culture into segregated interest groups."[97] In fact, Shattuck juxtaposes the passing on of an integrated culture as the primary purpose of schooling with a view of education proposed by "Americans long ago" such as "Jefferson, Horace Mann, and John Dewey," who decided that education was "the best vehicle . . . to change society"—that "free public schools could serve to establish a common democratic culture."[98] The kind of social change Shattuck envisions, however, is a form of cultural assimilation to forms of social and cultural hegemony for the purposes of adapting to existing social conditions. It is not about challenging abusive forms of power in the interests of social transformation, as Jefferson requires. Within Shattuck's rhetoric, the Jeffersonian view of civic education as a means for preserving individual rights and property and for enabling non-repressive self-government gets rearticulated as a deeply divisive, politicized mechanism that teaches "propaganda and advocacy" in the service of special interests such as "minority groups, feminists, gays and lesbians, Marxists, and the like."[99] Here Shattuck capitalizes on a mainstream logic since the Reagan era, which suggests that politics in general, or what is now commonly referred to as "Big Government," uniformly works to protect special interest groups at the expense of individual, taxpaying (read: white) citizens. Though this is not what the author meant by "candor and perversion," such a recoding of civic education demands forceful engagement and challenge.

Perhaps most interesting for our purposes is Shattuck's reliance on biological evidence to forward his racist arguments against the Jeffersonian model of the political education of citizens. Confronting the problem of the purpose of higher education as either a means for socialization within an existing culture or for "challeng[ing] and overthrow[ing] that culture," Shattuck appeals to the following analogy between human biological reproduction and education. After fertilization, "the human embryo [apparently a metaphor for the college student] sets aside a few cells . . . sheltered from the rest of the organism."[100] These, we are reminded, have the special ability to reproduce sexually. "Our gonads," Shattuck continues, "represent the most stable and protected element in the body and are usually able to pass on unchanged to the next generation the genetic material we were born with." Thus "the sins of the fathers and mothers ... are not visited upon their children."[101] As no such biological process exists in cultures, the analogy continues, all cultures have nonetheless devised something similar— what "we call *education*. By education, we pass on to the young the customs, restrictions, discoveries, and wisdom that have afforded survival so far."[102] And thus Shattuck draws the following conclusion: "There is good reason to maintain that, unlike other institutions—political, social, and artistic—which may criticize and rebel against the status quo, education should remain primarily a conservative institution, *like our gonads*."[103] While we see no need to deconstruct what one might call Shattuck's "gonad theory of education," it is interesting to note that such appeals to biology—we might recall here Alfred Schulz's shameless greyhound analogy— are always on the side of the dominant order. Or more precisely, biologism is waged against politics itself, as natural law is repeatedly invoked to sanction racial inequality and exclusion. Thus it is imperative to challenge, as we've attempted throughout this chapter, the deeply antipolitical and racist sentiments that scholars such as Shattuck give voice to. Such commentaries, though ludicrous at times, nonetheless resonate with broader public discourses that are fundamentally an attack on political democracy—either through an assault on public spaces for deliberation and dissent like the university or on the notion of difference itself.

Conclusion

Recent progressive work in rhetorical and cultural theory takes aim primarily at notions of citizenship that denigrate individual and collective agency and forms of civic education that reinvent racist national traditions rather than expand the scope of individual freedoms and the conditions for democratic public life. In short, what such critics share is a commitment to education as, in Paulo Freire's words, an ethical and political act of "intervention in the world."[104] Such a commitment, we have tried to demonstrate, is entirely in keeping with the historical responsibilities of the university, as Thomas Jefferson and others conceived it, to produce an active and critical citizenry. But as the above debates indicate, citizenship and civic education are highly and historically contested terms. Just as there is nothing self-evident (in spite of their rhetoric) about the largely ascriptive notion of citizenship that Hart, Hirsch, and others subscribe to, there is nothing self-evident about their concept of an appropriate college curriculum for producing good citizens. We have attempted to show that the very historical moment when the conception of citizen as bearer and protector of Anglo-American cultural traditions displaces the liberal-republican citizen as bearer of rights and duties is also the moment when the liberal arts curriculum shifts from classical rhetoric to literary studies and the subsequent racist invention of national cultural tradition. We have also tried to demonstrate, after Raymond Williams, how the pedagogical imperative of higher learning correspondingly shifts from the production of texts to their consumption, and from production of active citizens to passive consumers of high culture. Although we have been largely concerned with mapping the historical conditions—inflected by the politics of race—that led to these transformations, our purpose has been to demonstrate just how central race is to any understanding of past and present notions of citizenship and civic education and their relationship to the liberal arts. Just as any call to rethink notions of citizenship and civic education must consider a history of racialized exclusions in the United States, so too must it engage the problem of political agency in a neoliberal era. We will examine these questions more specifically in later chapters.

Chapter 5

The Return of the Ivory Tower: Black Educational Exclusion in the Post–Civil Rights Era

The year 2003 marks the one-hundredth anniversary of W. E. B. Du Bois's most celebrated publication, *The Souls of Black Folk*. An astonishing work of literary, historical, and sociological merit, *Souls* has inspired generations of academicians and activists alike drawn to the politics of identity, the color line, double consciousness, the talented tenth, and theories of race. The year also marks the fortieth anniversary of Du Bois's death in Ghana in August 1963. Save for *Souls*, the memory of Du Bois appears perpetually in danger of passing into oblivion, given the concerted efforts on the part of established powers to radically curtail his contributions to twentieth-century social and political thought. It seems that the commemoration of *Souls*—written by Du Bois in his early thirties—sanctions an official burial of the next 60 years of the author's life, which were devoted to scholarly examination of and struggle against racist exploitation and exclusion at home and U.S. imperialism and colonialism abroad. Though his relationship to the university was strained, his commitment to education as a primary mechanism for individual self-determination and collective democratization never wavered. Even as he explored the transformative potential of more public sites of pedagogy, he continued to

produce groundbreaking studies in urban sociology, histories of
the transatlantic slave trade, and his monumental *Black
Reconstruction in America*, as well as several now-classic works in
the field of education and numerous works of poetry and fiction
(including five novels). An engaged public intellectual, Du Bois was
a founder of activist organizations like the National Association for
the Advancement of Colored People (NAACP) and an editor of its
famed magazine, *The Crisis*. The widely acknowledged "father" of
the Pan-African movement, he also helped to organize and lead the
successive Pan-African congresses in efforts to forge an international coalition united in the pursuit of racial justice and human
rights. He was particularly vocal about the collusion of the
U.S. government with colonial empires under the auspices of fighting off the communist menace and making the world "safe for
democracy," a form of red-baiting, he argued, largely designed to
silence those who privileged peace, human rights, and democracy
over corporate interests.

For his courageous opposition, Du Bois was rewarded with endless harassment and overt repression. During his own lifetime, both
the academy and the federal government sought to minimize his
political and pedagogical influence. His relationship with institutions
of higher learning always a stormy and ambivalent one, Du Bois felt
thwarted by successive universities—from his early years at
Wilberforce through his tenure at Atlanta University—in his efforts
to link rigorous scholarship with progressive social change to make
the university a genuinely democratic public sphere. Moreover, the
government brought criminal charges against Du Bois and restricted
his travel to foreign countries eager to receive him, successfully
containing for a time his efforts to construct a global public sphere
organized against Cold War militarism, racism, colonialism, and
economic inequality.

As the university's civic mission is imperiled by corporatization
and racial backlash, access to its resources are increasingly predicated on whiteness and wealth, and the greater public good is
financially and spiritually starved with every advance of American
empire abroad, now is surely the time to recall the legacy of
Du Bois, as both scholar and activist. In what follows, we want to
develop this line of argument by first exploring the pedagogical

implications of Du Bois's work outside his specific contributions to the field of education, beyond the now ritual engagement with notions of "the talented tenth" and its later revisions. His engagement with historic struggles for racial justice and democracy, we argue, has much to teach contemporary progressives in and out of the university attempting to challenge the antidemocratic excesses of a far-right federal government and its drive toward empire. Second, drawing on Du Bois's insistence that formal education is central to the functioning of a nonrepressive and inclusive polity, we want in this chapter to reflect further on the current crisis of schooling at all levels, which must be taken up within the context of neoliberal social and economic policies as well as the racist backlash against the civil rights gains of the 1960s. Finally, we will address the degree to which engaged dialogue about the history and politics of racialized exclusion in the United States and globally in the university have been derailed by the dictates of a particularly limp version of liberal multiculturalism and its allegiance to the privatized discourses of identity and difference. In so doing, we will explore, in a partial and incomplete way, the role that educators might play in linking rigorous scholarship and critical pedagogy to progressive struggles for securing the very conditions for racial justice and political democracy.

To be sure, the achievements of *Souls* are indeed vast, but it may not be Du Bois's most important or relevant work for our time. That distinction, we argue, belongs to his magnum opus, *Black Reconstruction in America*, which he completed in 1935 at the mature age of 68. A lesser-known text of more than 700 pages, *Black Reconstruction* has the artistic refinements of *Souls*, though its political, philosophical, and historiographic reach and insight remain unsurpassed. In these pages Du Bois offers a more fully developed examination of the interconnections between political economy and racial oppression, and between forms of state terrorism wielded against a domestic population in the name of "law and order" and those exported globally, dignified under the rubric of "foreign policy." Not only did Du Bois rewrite the conventional and deeply racist interpretations of Reconstruction as a failed project by reigning historians of the day, but he revised basic tenets of the philosophy of history held since the Enlightenment. In his account, Reconstruction was an era of unprecedented civil rights victories, as black Americans achieved the rights of full citizenship; voted in elections; held political office in local, state, and federal government; and established free public education in the South, as well as dozens of black colleges where none had existed before. In short, it was a period of democratic rebirth in the wake of the Civil War when there was hope that the nation, now unified in freedom, was taking its first steps toward the ideals of its Constitution. But that hope was short-lived; Du Bois recounts the swift and pervasive counterrevolutionary response to that period's democratic advance, which gained national momentum in 1876 with the election of Rutherford B. Hayes to the presidency and culminated in the *Plessy v. Ferguson* decision of 1896. The net result of two decades of reaction was to push black America "back toward slavery." Writing in the midst of the Great Depression, in the thick of the Jim Crow South (he had returned to Atlanta University after his resignation from the NAACP), Du Bois rightly challenged the presumptive "forward march" of history.

Indeed, it is ironic that Du Bois witnessed in 1935 Depression-era poverty crushing southern blacks and the collapse of the labor party while writing about the economic crisis of the mid-1890s, the defeat of the agrarian Populists, and the establishment of Jim Crow. The period must have seemed to him to repeat uncannily the Long Depression of the 1870s and the disciplining of labor by the new industrial capitalists who easily manipulated postwar Reconstruction efforts to serve their interests, and just as shamelessly abandoned them when they didn't. How did he make sense of the complexities of such vertiginous cycles of democratic advance and decline? Du Bois's answer, in part, was to look to the color line, "the Blindspot of American political and social development," that crippled the scholarly search for a truthful and coherent history and "made logical argument almost impossible."[1] Du Bois understood the ill-fated history of Reconstruction as part and parcel of the ongoing tragedy of American political life—a tragedy to beggar the Greek—which resides in the nation's failure to grasp that the "problem of race" involves the very foundations of American democracy, both political and economic.

While enduring economic crisis and racial repression in Atlanta, Du Bois pondered the capacities of southern politicians and northern business leaders (strange bedfellows indeed!) to secure the unity of interests that otherwise stood worlds apart. *Black Reconstruction* sought to explain, in other words, the loyalty of the white laborer not to the black laborer, but to a dethroned southern aristocracy and the eventual obeisance of both to the will of northern industrialists in the decades following emancipation. With characteristic grace and insight, he wrote:

> Thus by singular coincidence and for a moment, for the few years of an eternal second in a cycle of a thousand years, the orbits of two widely and utterly dissimilar economic systems coincided and the result was a revolution so vast and portentous that few minds ever fully conceived it; for the systems were these: first, that of a democracy which should by universal suffrage establish a dictatorship of the proletariat ending in industrial democracy; and the other, a system by which a little knot of masterful men would so organize capitalism as to bring under their control the natural resources, wealth and industry of a vast and rich country and through that, of the world. For a second, for a pulse of time, these orbits crossed and coincided, but their central suns were a thousand light years apart, even though the blind and ignorant fury of the South and the complacent Philistinism of the North saw them as one.[2]

His substantial and meticulously detailed investigation exposed the uses of racial resentment, inflamed by antiblack ideologies circulating as science, religion, or political theory, to align these vastly opposed "orbits," thus securing the hegemony of powerful business interests as it consolidated white racial rule.

Though his conceptual framework has been dismissed as so much Marxist dogma applied to history, the criticism seems to miss the central contributions of *Black Reconstruction* to contemporary political thought: that in the pursuit of genuine democratization and an equitable distribution of resources, questions of race cannot be meaningfully separated from questions of class. Furthermore, struggles for freedom and social justice are necessarily global in their effects, if not in their reach. The untoward abuses of black workers that followed quickly on the heels of their manumission had consequences as well for their white working-class counterparts. To the degree that such unrestricted profiteering enabled American industrialists to advance on global markets, its effects were felt worldwide. In short, the continued allegiance to racist ideologies and exclusions undermined not only black emancipation in the United States but the emancipation of humankind more generally. Anticipating the vulnerability and insecurity of contemporary American workers in general (and black and brown in particular) occurring in tandem with the exploitation of nonwhite labor in the "developing" world as a result of Western deindustrialization, Du Bois's insight remains unimpeachable, even though his language, derived from nineteenth-century Marxist thought, is rather orthodox.

Du Bois argued that the Reconstruction era was not simply a battle between black and white races, or between master and ex-slave; it was also a vast labor movement galvanized by the promise of industrial democracy, which was to eventually betray itself on the alter of racial apartheid at home and imperialism abroad. And the implications of that failure of democratic vision and will were felt everywhere. As historians since Du Bois have convincingly argued, one might rightly question the recasting of Civil War struggles in terms of a general strike or even the desire for a "dictatorship of the proletariat." But the fact remains that by 1876, the dream of political democracy in the United States was deferred indefinitely. As a result of ensuing decades of unchecked corporate and imperial power, the nation and the world were eventually plunged into economic depression and war. The hopes of Reconstruction dashed, Du Bois pondered the violence and carnage that followed in its wake, not just for black Americans, but for ordinary citizens the world over. His eloquent assessment is worth repeating at length:

> God wept; but that mattered little to an unbelieving age; what mattered most was that the world wept and still is weeping and blind with tears and blood. For there began to rise in America in 1876 a new capitalism and a new enslavement of labor. Home labor in cultured lands, appeased and misled by a ballot whose power the

dictatorship of vast capital strictly curtailed, was bribed by high wage and political office to unite in an exploitation of white, yellow, brown and black labor, in lesser lands and "breeds without the law." Especially workers of the New World, folks who were American and for whom America was, became ashamed of their destiny. Sons of ditch-diggers aspired to be spawn of bastard kings and thieving aristocrats rather than of rough-handed children of dirt and toil. The immense profit from this new exploitation and world wide commerce enabled a guild of millionaires to engage the greatest engineers, the wisest men of science, as well as pay high wage to the more intelligent labor and at the same time to have enough surplus to make more thorough dictatorship of capital over the state and over the popular vote, not only in Europe and America but in Asia and Africa.

The world wept because within the exploiting group of New World masters, greed and jealousy became so fierce that they fought for trade and markets and materials and slaves all over the world until at last in 1914 the world flamed in war. The fantastic structure fell, leaving grotesque Profits and Poverty, Plenty and Starvation, Empire and Democracy, staring at each other across World Depression. And the rebuilding, whether it comes now or a century later, will and must go back to the basic principles of Reconstruction in the United States during 1867–1876.[3]

Du Bois's depiction of the age of rampant industrialization, robber barons, and racist exploitation thus challenges a form of contemporary common sense that equates a free market economy with democracy by invoking the power of capital to denigrate those values Americans hold most sacred—freedom, self-reliance, and a level playing field, as well as the maintenance and protection of family and community. The unfettered power of capital rendered free elections relatively meaningless through graft, corruption and the unequal distribution of power and resources (or, in today's terms, through the corporate media's near monopoly control over people's access to information and critical viewpoints). Unfettered capital also undid the bonds of family and community to the degree that it fostered identifications with the powerful and wealthy, however distant or remote, and a corresponding indifference to the plight of everyday people. But Du Bois's most startling insight is his prescient understanding of the global consequences of economic expansion and white racial domination in the interests of "nation building." Images of "grotesque Profits and Poverty, Plenty and Starvation, Empire and Democracy" will no doubt resonate powerfully, if not eerily, for the twenty-first-century reader. As U.S.-led global capitalism advances its neoliberal economic and social agenda, pauperizing vast nonwhite populations of the world, the clear majority of U.S. citizens experience downward wage pressure and a rapidly declining standard of living, bearing the expense materially and psychologically as the military arm of the world's remaining superpower enters into a permanent war on terrorism to protect such "freedoms." Though there is growing recognition of the role that oil plays in the current U.S. military occupation of Iraq, precious few intellectuals link historically racist Western attitudes toward Arabs to the "war on terror," preferring instead a more coded rhetoric of religious and civilizational conflict. An awareness of the forces at work in the global movement of capital and the commitment to racial domination and exclusion—separate forces, yet inseparable—seems to lie just beyond the consciousness of average American citizens, though the consequences of these abstract pressures insinuate themselves into nearly every aspect of their lives.

Du Bois's *Black Reconstruction*, however, is not written to workers black and white, then or now, who have been ongoing victims of rapacious capital or racist ideological subterfuge. It is preeminently a challenge to the chroniclers of official history, to intellectuals and academics who recklessly use "a version of historic fact in order to influence and educate the new generation along the way [they] wish."[4] In the final chapter of *Black Reconstruction*, Du Bois confronts those historians who, assuming as axiomatic the inferiority of the "Negro," participated in the "most stupendous" campaign "the world has ever seen to discredit human beings" in the interests of economic and racial domination, an effort involving the pedagogical force of the entire culture, of "universities, history, science, social life and religion."[5] The sting of this indictment is surely felt today among progressives who wonder how the United States got so far afield of its reputed values of liberty, equality, and justice for all. And so the gauntlet is thrown down to intellectuals who desire a more democratic future than the dystopic present offers, as Du Bois grasped that the revitalization of democratic politics is preeminently a pedagogical endeavor. Education in the university, or informally circulating through earnest public debate, may not be all that is required to counter the antidemocratic excesses of empire's most zealous advocates, but its absence guarantees the brilliant success of radically antidemocratic agendas. As we take the measure of the new American empire it seems appropriate to pose, after Du Bois, the question of the color line, in an effort to avoid what Claude Lemert recently termed, with not a little chagrin, "pastism," or "the error of failing to grant the real or virtual dead credit for having understood present matters better than present company . . . a refusal to grant that others knew the rules before [we] did."[6]

Looking Backward: Post–Civil Rights and Post-Reconstruction (2003–1876)

Du Bois was uncannily accurate in his prediction that the rebuilding of American political democracy "whether it comes now or a century later, will and must go back to the basic principles of Reconstruction in the United States during 1867–1876." Nearly a century after its initial attempts at black emancipation, a renewed Civil Rights movement attempted to rebuild a democratic society out of the ashes of Jim Crow. Throughout the decade of the 1960s, organized citizens agitated for and achieved the formal, legal repeal of segregation, the reenfranchisment of black voters, the passage of legislation guaranteeing equal opportunity in school and the workplace, and welfare benefits for black families. Yet the full realization of these rights and protections was not forthcoming. On the heels of such long-awaited victories, the nation's citizenry was told that civil rights had simply gone too far. In the ensuing decades, successive Supreme Court rulings have quietly dismantled *Brown v. Board of Education of Topeka*, creating rapidly resegregated and unequal schools reminiscent of the *Plessy v. Ferguson* era. Attacks on affirmative action have similarly diminished black student access to universities, as well as unleveled the playing field for black-owned businesses. Felony disenfranchisement, ongoing voter intimidation in the South, and the massive fraud perpetuated against black voters in the 2000 presidential election has rendered formal voting rights a partial accomplishment at best.[7] In the late 1960s, black women achieved the right to receive welfare benefits, but this, too, proved a pyrrhic victory. By the 1980s they had become the unwitting poster children for a propaganda campaign to dismantle the welfare state, as the federal government reneged on the provision of compassionate services and redoubled its repressive functions. In much the same way that historians, intellectuals, journalists, and others demonized the period of American Reconstruction, the political revolution of the 1960s that secured civil rights for blacks, women, the criminally accused, senior citizens, and others was immediately reframed in terms of a cultural revolution emphasizing permissiveness, lack of work ethic, moral relativism, and utter contempt for mainstream values.

The history of American apartheid and its ongoing effects disappeared from public memory in this proudly "post–civil rights era." We have once again entered an Orwellian era in American life, a time in which, as Du Bois once remarked, "logical argument [is] all but impossible." The general presumption of the post–civil rights era, roughly from mid-1970s to the present, is that racial injury and injustice derived from centuries of enslavement, Jim Crow, and urban ghettoization have been both widely acknowledged and corrected by established powers, and that we, as a nation, have transcended race in the interests of colorblind public policy. What is left out of that reading of contemporary political culture, of course, is the enormity of the backlash that ensued in the wake of civil rights advancements. As in the post-Reconstruction days, the terms of political discourse have been utterly subverted by conservative ideologues and pundits softpeddling racial reaction in various guises and for various purposes. From crude appeals to deviant black sexuality, criminality, and intellectual/moral inferiority (as in George Bush, Sr.'s "Horton" campaign ad or the vogue of Charles Murray in the mid-1990s) to more coded appeals to "colorblindness," and "race-transcendence," right wing intellectuals and experts from a variety of cultural spheres—corporations, think tanks, churches, universities, and especially the mass media—have successfully revived the antiblack rhetoric and imagery of a century ago. For neoconservatives, the goal of the post–civil rights era has been to roll back the gains made by social movements in the Sixties and radically restrict immigration in the interests of cultural nationalism and the consolidation of white political and economic power, while silencing any discussion of race in mainstream national politics by insisting on colorblind public policy. For neoliberals, it has been to "free" the markets through privatization and deregulation of formerly public goods and services provided by the state. To garner support for such a nakedly corporate agenda, they have also adopted a populist platform to "free" citizens of an oppressive tax burden by providing huge tax benefits to the wealthy and dismantling the welfare state (now rendered an archaic set of bureaucracies that largely fail the populations they are designed to serve by absorbing tax dollars to promote the dependency and debauchery of poor minorities) thus dissolving the language of public life and common good.

Having either bought into dominant ideologies or simply been denied access to any other, everyday people whose jobs are increasingly vulnerable in a rapidly globalizing economy and whose safety is ever more uncertain in a post-9/11 world have tied their interests to a party dominated by corporate and imperial interests—the very elite responsible for sending jobs and factories (and now soldiers and munitions) overseas to promote a more flexible workforce, secure scarce resources, and exempt themselves from all forms of accountability. Of the many achievements of this ongoing right-wing campaign, the preeminent victory has been the near elimination of frank and reasonable discussion of what is happening in our communities, our country, and across the globe. It is a victory as much determined by the right's tight control over national political rhetoric as it is by their capacity to manage and contain public spaces of pedagogy—both informally in the media[8] and formally through successive attacks on public education at all levels.

Having endured over two decades of conservative hegemony, some progressives in the United States are beginning to rethink the public role of pedagogy as they attempt to arouse the citizenry in efforts to "Take Back America" from the apostles of free market fundamentalism with their own counter-narratives of life in the new millennium. Missing from most analyses, however, is a sophisticated treatment of racial politics in the post–civil rights era, particularly as it intersects with the interests of neoliberal and neoconservative policymakers. Yet, if we are to learn anything from Du Bois's life and work, it is surely that strategies for political and economic democratization require sophisticated analyses of how race and class politics—though discrete discourses—inform and influence each other both within the United States and on the global stage. Before we elaborate on the significance of *Black Reconstruction* for understanding the current political context, we will provide brief examples of three exemplary, though incomplete, efforts to challenge the emergence of American empire.

In a May 2003 issue of *The Nation*, William Greider maps the "grand ambition" of the conservative forces guiding the second Bush administration. In a lead article entitled "Rolling Back the Twentieth Century," he asserts that the Right's primary objective is to recast the federal government under Bush, Jr. in the likeness of what it was—quite literally—under William McKinley, who held the office of president from 1897 to 1901. This was the gilded era of government when corporations and religious organizations reigned supreme, before any Progressive era New Deal, before, as conservative tactician Grover Norquist put it, the ascendancy of "Teddy Roosevelt, when the socialists took over."[9] For Norquist and the other conservative groups he has so masterfully organized in the last two decades, regaining paradise lost requires the demise of "big government." Though it has proven a popular catchphrase among mainstream Americans, Greider observes, the conservatives' objectives are quite radical by any standards. They are, in short, the elimination of federal taxation of private capital; phasing out of pension-fund retirement systems; the withdrawal of government from any direct role in housing, health care, and assistance to the poor; restoring the centrality of church, family, and private education to the nation's cultural life; strengthening business and market-based solutions to public concerns in areas like the environment; and, finally, dismantling organized labor. Ironically, while the enemy for conservative forces is "big government," the size of government has been vastly expanded under conservative rule in the last two decades. As Kevin Baker pointed out in *Harper's*

Magazine, "since the advent of Reagan and the current Republican hegemony the federal government has by almost all objective measures become larger, more intrusive, more coercive, less accountable, and deeply indebted than ever before. It has more weapons, more soldiers, more police, more spies, more prisons."[10]

Similarly, in his opening speech at the "Take Back America" conference sponsored by the Campaign for America's Future a month later, Bill Moyers discussed the influence of the McKinley administration, and especially Mark Hanna, on Karl Rove, Bush's reputed brain. Hanna was the primary architect of McKinley's public persona and largely responsible for his successful bid for the governorship of Ohio and later the presidency—much as Rove enabled Bush's ascendancy in Texas and his transition to the White House. The political achievements of Mark Hanna, Karl Rove's hero, were largely the result of old-fashioned corporate shake-downs. Hanna, Moyers asserts, "saw to it that first Ohio then Washington were 'ruled by business . . . by bankers, railroads and public utility corporations.'" This "degenerate and unlovely age," Moyers notes, is the "seminal age of inspiration for the politics and governance of America today."[11]

Then in the August 2003 issue of *Harper's*, Lewis Lapham offered yet another scathing indictment of present-day chicanery by invoking days of future past. Lapham examines McKinley-era weapons of mass deception used to preempt the threat of an emergent Populist movement for social and political reform in the face of widespread economic inequality. "If by 1890 the Industrial Revolution had made America rich," Lapham writes, "so also it had alerted the electorate to the unequal division of the spoils. People had begun to notice the loaded dice in the hand of the rail-road and banking monopolies, the tax burden shifted from capital to labor."[12] Reminiscent of patterns of instability in so-called boom cycles like the 1990s, economic depression and widespread unemployment in the winter of 1893–1894 aroused the nation's citizenry, and the government literally reached for its guns. Looking for "'something' in the words of an alarmed U.S. senator, to knock the 'pus' out of this 'anarchistic, socialistic and populist boil,'" Lapham continues, "the McKinley Administration came up with the war in Cuba, the conquest of the Philippines, the annexation of

Puerto Rico, and an imperialist foreign policy. . . . Only by infecting the republic with the delusion of imperial grandeur could the nation . . . smother the republican spirit and replace the love of liberty with the love of the flag . . . [as] all political quarrels [were] suspended in the interests of 'the national security.'"[13] Moving between past and present, between steel interests and oil interests, a war in Cuba and a war in Iraq, Lapham's indictment of the Bush administration and its propaganda machine is penetratingly clear. It would be a grave mistake, of course, to assume that U.S. military operations in the Middle East were merely a means to distract (or silence) American citizens, though they may well function in this way, rather than as part of a broader effort to secure resources, labor, and trade advantages from formerly colonized nations. What Lapham's analysis gestures toward is an opportunity to gauge the impact of unfettered corporate and imperial power on civilian populations both at home and abroad—and a crucial opportunity to challenge the racism that promotes complicity with such abuses.

If the project to "take back America" in the interests of substantive, participatory democracy has a chance to take root among everyday citizens, it is precisely because of committed intellectuals working in the public interest of the caliber of Moyers, Greider, and Lapham, whose acumen and courage in the face of a rising tide of jingoistic patriotism and relentless merging of government with corporate power have been unflinching. What remains under the radar, however, even among the most outspoken proponents of democracy today, is the "preemptive" war at home on Black America, the domestic corollary to an expanding "nation building" agenda abroad. Lost from these perspectives is both the legitimacy of race as a political issue and the ongoing racism that eats away at the moral condition and the very foundation of democracy. That the current Bush administration emulates the McKinley era for its show of raw corporate and imperial power is surely correct. But the graphic juxtaposition of two eras of unprecedented political corruption explains the ascendancy of neither. Hence it is worth investigating another parallel to that brutal era left unexamined, what Du Bois called "the American blindspot." If we are to draw any meaningful conclusions from turn-of-the-twentieth-century hijacking of political democracy, we must expand our understanding of the racial politics of that era and our own, as they are inextricably entwined with corporate and imperial agendas.

We would like to advance the argument, therefore, that the McKinley administration represents the culmination of a series of events set in motion in the aftermath of the Civil War and southern Reconstruction, just as the current Bush administration reflects the organizational pinnacle of right-wing reaction to the civil rights revolution of the 1960s. The revolution of 1863–1876 and the revolution that spanned the decade of the 1960s won (for a time) first for black men, then for both sexes the rights, responsibilities, and protections of democratic citizenship, at least in theory. Among the rights afforded citizens were those that vouchsafed the capacity for self-possession and self-determination—the right to paid labor, education, enfranchisement, as well as protections from state violence in its various forms—insult, humiliation, isolation, incarceration, starvation, disease, and murder. On the heels of each revolutionary victory, however, came a swift and pervasive counterrevolutionary response, and as we will later detail, education would prove a primary battlefield throughout.

Following the southern Reconstruction experiment, the national Republican party of the late 1800s, the party of Abraham Lincoln, eventually abandoned its efforts to achieve racial equality after it failed to win the support of southern conservatives and faced potential political defeat. Rutherford B. Hayes, for example, secured his quite controversial election victory in 1876 by promising to stop enforcing civil rights and promptly withdraw federal troops from the South. The concessions to former Confederates came fast and furious, and all at the expense of newly manumitted slaves. By the year 1896, McKinley won the presidential election and "separate but equal" was rendered legal by the landmark *Plessy v. Ferguson* decision. Not only did black Americans suffer the formal loss of civil rights and legal protections as a result of the Supreme Court's ruling (most had already experienced the actual loss of these in the two decades leading to the decision), but they were increasingly subject to forms of domestic terrorism throughout the 1890s, with some 200 lynchings occurring on average per year. A time of cynicism and deep despair among black intellectual leadership, it was an era marked by the ascendancy of black leaders like Booker T. Washington, who, in concert with majority white opinion, rejected civil rights struggles in favor of philanthropy and self-help.

Facing its own political crisis nearly 100 years later, the contemporary advocates of civil rights, the Democrats, took note. After several Republican presidential victories from 1968 to 1988, elections won by a virtually all-white party utterly opposed to civil rights, the Democrats followed suit and excised the rhetoric of racial justice from its national platform. The Democratic Leadership Council (DLC) was formed in the mid-1980s by mostly white, male, largely southern Democratic politicians, corporate lobbyists, and fundraisers to counter the "liberal fundamentalism" of the party's base—principally blacks and other minorities, unionists, feminists, and Greens. The DLC quickly gained ascendancy on a platform (backed by corporate dollars) that the Democratic party had become too solicitous of African American and Latino political support, too respectful of workers' rights, and too responsive to the peace, justice, and environmental movements.[14] The success of the DLC culminated in the election of Bill Clinton, the DLC/New Democrat candidate for president in 1992, a triumph that spelled a political disaster for black constituencies. For example, Bruce Dixon observes, "Rather than answer the Reaganite myth of the welfare queen, Clinton pandered to it and gave us a 'welfare reform' more punitive than anything Reagan-era Republicans could have wrested from Congress."[15] To assuage African Americans in light of such stark failures to serve that constituency, a new class of "black Trojan horse Democrats" appeared, who, financed by corporate power, are being foisted on black communities like so many latter-day Booker T. Washingtons to allegedly "represent" their interests.

As the unprecedented corporate influence on the political systems of each period makes clear, the post-Reconstruction era and the more recent post–civil rights era were periods of unprecedented economic change enabling the rapid consolidation of corporate power. The former began with the Long Depression that commenced with the panic of 1873. It was the bust that ended what Eric Hobsbawm called the "Age of Capital," the largest period of economic expansion in the early history of capitalism.

Following a series of wage cuts among the nation's railway workers, who already suffered low wages, dangerous working conditions, and the scheming and profiteering of railroad companies, the year 1877 erupted in a series of tumultuous strikes extending from cities in the Northeast to St. Louis. As a result, the nation experienced a run on banks and the failure of thousands of businesses, followed by the collapse of half the nation's iron producers, and half its railroads between 1873 and 1878, leading to the failure of other business and industries that had risen in tandem with heavy industrialization. While some concessions were made to working Americans, northern industrialists strengthened their position in collusion with southern planters. In addition, revolutionary technological breakthroughs increasingly made forms of manual labor redundant, as innovations in steam and electricity replaced human muscle. By the year 1877, the historian Howard Zinn writes:

> The signals were given for the rest of the century: the black would be put back; the strikes of white workers would not be tolerated; the industrial and political elites of North and South would take hold of the country and organize the greatest march of economic growth in human history. They would do it with the aid of, and at the expense of, black labor, white labor, Chinese labor, European immigrant labor, female labor, rewarding them differently by race, sex, national origin, and social class, in such a way as to create separate levels of oppression—a skillful terracing to stabilize the pyramid of wealth.[16]

Although it is true that many black Americans have come to enjoy middle-class status in the post–civil rights era, their share of the economic pie since the days of Reconstruction has improved only slightly. Consider that in 1865, blacks owned 0.5 percent of the nation's net worth, and in 1990, their net worth totaled only 1 percent.[17] With the advent of postindustrialism, the United States witnessed the frenzy of breakneck deindustrialization, the weakening of organized labor, increasing joblessness, poverty, and decaying infrastructure particularly in the nation's cities, as well as the attendant psychological fallout of fear, anxiety, and insecurity. Following increased competition from newly rebuilt European and Japanese industries, whose urban centers were all but decimated in

World War II, as well as a series of wildcat strikes at home in the
early 1970s, U.S. industries began to take flight—enabled by startling advances in transportation and later in information technologies. In the next few decades, the nation would experience the loss
of millions of jobs in manufacturing as well as a significant decline
in wages as corporations moved factories overseas to developing
nations offering cheap, nonunionized labor, totally deregulated
lands beyond the reach of any environmental protections groups,
and tax holidays—as well as the chance to send a resounding message to American workers. While blue-collar workers in general
suffered from the flight of industry overseas in the last 25 years,
black Americans were particularly squeezed, as historians and soci-
ologists like Manning Marable and Howard Winant have extensively documented.[18] Because pay for blacks historically has been
higher in manufacturing than in many other fields, deindustrialization has hit blacks disproportionately harder than whites. Since
2001, for example, the United States has lost 2.6 million jobs;
nearly 90 percent of them were in manufacturing. According to
Louis Uchitelle of the *New York Times*, "In 2000, there were two
million black Americans working in factory jobs, or 10.1 percent
of the nation's total of 20 million manufacturing workers." Since
the recession began in March 2001, "300,000 factory jobs held by
blacks, or 15 percent, have disappeared. White workers lost many
factory jobs, too—1.7 million in all. But because they were much
more numerous to begin with, proportionally the damage was less,
just 10 percent."[19] While unemployment for black men has generally been double the rate for their white counterparts, 10.5 percent
according to one low estimate,[20] the jobless rate among minority
teens is the highest in 55 years. According to the Children's Defense
Fund, June 2003 jobless rates were "78.3% (the highest since
1983) for black teens and 68.4% for Latino teens, the highest
reported for young Latinos."[21] In 2003, corporate analysts predict
that the nation's service sector and information technology jobs
will quickly follow suit; "3.3 million U.S. service industry jobs and
$136 billion in wages will move offshore to countries like India,
Russia, China, and the Phillippines" in the next 15 years, which
will further stress an already tight labor market and continue to
drive wages down as the cost of health care and education continue
to spiral to outrageous heights.[22]

Given the untoward vulnerability of American workers in both eras, the pervasive mood of the post-emancipation era, as W. E. B. Du Bois describes in *Black Reconstruction*, is also that of our current situation: a pervasive and multivalent fear. The insecurity, anxiety, and uncertainty of citizens were aroused by many things, Du Bois wrote, "but usually losing their jobs, being declassed, degraded, or actually disgraced; of losing their hopes, their savings, their plans for their children; of the actual pangs of hunger, of dirt, of crime. And of all this, most ubiquitous in modern society is that fear of unemployment."[23] And fear is what propelled the mob violence and lawlessness that was in 1865–1868 "spasmodic and episodic," only to become organized and systemic throughout the South in the coming decades. "Lawlessness and violence filled the land," Du Bois reflected, "and terror stalked abroad by day, and it burned and murdered by night. The Southern states had actually relapsed into barbarism. . . . Armed guerrilla warfare killed thousands of Negroes; political riots were staged; their causes or occasions were always obscure, their results always certain: ten to one hundred times as many Negroes were killed as whites."[24] Today's culture of fear is, of course, the topic of much scholarly attention, spawning what Mike Davis recently called "fear studies." From sophisticated theoretical analyses such as Ulrick Beck's *World Risk Society* to Barry Glassner's more popular *Culture of Fear*, intellectuals have taken up the perils of postmodern society, both real and manufactured, for cynical exploitation and legitimation of lethal force. Politicians in both periods were able to successfully focus a generalized anxiety and uncertainty about unemployment, poverty, homelessness (and, in the new millennium, nuclear annihilation, terrorist attacks, and environmental devastation) into a specific, individualized fear of crime and personal concern over safety, as both periods witnessed the "official solidification of centuries-old association of blackness with criminality and devious violence."[25] As the result of the successful mobilization of racialized fears, the mob violence over a century ago has been supplanted by the prevalence of profiling, harassment, brutality, and even murder of black and brown populations (the high-profile cases of Rodney King, Amadou Diallo, Tyisha Miller, and Abner Louima offer ready examples) by an increasingly paramilitarized police force, now functioning as the legitimate arm of the law.

The generalized instability and anxiety of the post-Reconstruction and post–civil rights eras was thus the result of revolutionary changes in the economy no less than dramatic challenges to and transformation of social norms and political institutions that privileged whiteness, as black Americans demanded the rights and entitlements of full citizenship. In both instances, however, they found themselves on the bad end of public policies meant to bolster and appease a white citizenry discomforted by political, economic, and social upheaval. This is what Du Bois referred to as a "public and psychological wage" of whiteness to compensate for the decline in real wages, which translated into turn-of-the-century protocols requiring public deference and titles of courtesy extended to them because they were white—not to mention their admittance to public parks and beaches, attendance at the best schools, the right to vote, and so on. Black Americans, conversely, suffered the spiritual and material weight of the color line with the establishment of Jim Crow laws and social practices. The brief period after the Civil War in which black men voted, elected black representatives to the state legislatures and Congress, and introduced public schooling, at the state's expense, for all children in the South was effectively over. Free and equal participation in democratic life gave way to racial domination, ghettoization, segregation, and disenfranchisement.

Like the racial backlash that followed Reconstruction, the response to years of peaceful, nonviolent struggle for civil rights spanning the decade of the 1960s was an escalation of war rhetoric on the part of the established order, coupled with real destruction and casualties, as Johnson's War on Poverty gave way to Nixon's War on Crime and then Reagan's War on Drugs, which has been expanded and folded into George Bush, Jr.'s War on Terrorism. As the stagnation of wages, growing unemployment, and insufficient safety nets only deepened poverty for Americans on the low end of the wage spectrum in the 1970s and 1980s, the field of wartime operations shifted from the conditions of impoverishment to the poor themselves, who were increasingly cast as a racialized and disposable population. Efforts to polarize the electorate along racial lines picked up speed with Nixon's "tax revolt" of the late 1970s, which pitted taxpayers against "tax recipients" and fueled a dramatic attack on the welfare state. Reagan's 1980 campaign capitalized on this polarization, attacking programs from affirmative action to food stamps, and extending its reach to a rejection of any government interference in the economy and an all-out assault on "big government." A nightly battalion of conservative social scientists, legal scholars, educators, and preachers endowed with media omnipresence waged an ideological war against the "underclass," a term used to criminalize the poor by transforming them into a tangle of filth and human garbage. Their mantra: big government handouts had corrupted black communities by creating generations of cheats characterized by laziness, drug addiction, sexual excess, and a general taste for criminality and violence. As a result of such rhetoric, white working-class and middle-class voters increasingly perceived the Democrats' civil rights agenda to be in the service of blacks as well as feminists, gays, and other marginalized groups. "Quotas," "preferential treatment," and "groups" were so many code words used by the Reagan administration to signal to largely white suburban voters that the era of big government handouts to minorities was now over.[26]

Whereas turn-of-the-century recompense to white workers turned on appeals to white supremacy and the real material privileges of whiteness under Jim Crow, more recent attempts to assuage the pain of white workers in the new economy came in the form of public recognition of, and vows to end, their alleged victimization by big government policies favoring racial set-asides and affirmative action—or in the parlance of conservatives, "reverse racism," now challenged in the name of "fairness" and "colorblind" public policy. Big government critics at that time were indifferent to the irony of how liberal social programs were completely dwarfed by mushrooming military spending (just as today). "During the Reagan–Bush years," John Brenkman notes, "working class and middle class whites were willing to accept the massive shift of wealth from the middle class to the rich so long as they simultaneously perceived that Reagan's policies were transferring wealth *from blacks to whites*."[27] By the 1992 election year, it became clear to Democratic party officials that they could not win an election without wooing back the so-called Reagan Democrats who had fled the party in the previous decade. In *Chain Reaction*, an incisive analysis of the impact of Reagan–Bush politics on the Democratic Party, Thomas and Mary Edsall argue that the Clinton campaign appropriated the slogans of the Reagan–Bush era and crafted an explicitly "race-neutral" platform, which "voters were known to interpret in strictly racial terms."[28] What cannot be overemphasized here is the role that race has played in abetting both conservative and neoliberal efforts to depoliticize popular constituencies and privatize all remaining public goods and services through the discourses of anti-statism and self-help. Repressive state institutions—juvenile justice, the police, the prison—were the only institutions left to tend to the poor.

As the history of post-Reconstruction and post–civil rights reveals, there is a very short distance between blaming oppressed or excluded groups for their own misery, demonizing them for their poverty, and then criminalizing their behaviors. All over the South, the passage of Black Codes in the years after Reconstruction enabled the mass incarceration of former slaves who committed "crimes" such as vagrancy, absence from work, ownership of firearms, or violations of racial etiquette—practices that in other words were quite legal if one were white. It is worth noting that during slavery there were no blacks in prison; punishment for any transgression was meted out by the master of the plantation. "On the morrow of Emancipation, Loïc Wacquant notes, "southern prisons turned black overnight. . . . The introduction of convict leasing as a response to the moral panic of crime presented the double advantage of generating prodigious funds for the state coffers and furnishing abundant bound labor."[29] It was for these reasons that Frederick Douglass referred to incarcerated blacks as "prisoners of war," rather than criminals.[30]

Over 100 years later, Douglass's insight still has teeth. In the midst of racial backlash in an allegedly "colorblind" era, the war on crime and the war on drugs shook loose from their metaphorical moorings and became a real war. The renewed interest in prison labor, prison privatization, and what Paul Street has called "correctional Keynesianism"[31] are the contemporary corollaries of earlier efforts to both contain and extract free labor from a potentially subversive, largely black prison population. Politicians looking to outdo each other on "get tough" crime policies militarized city spaces, armed police with paramilitary weaponry and surveillance equipment, and made lifers out of nonviolent offenders with "three strikes" laws, expanding what critics have called the prison-industrial complex. In the span of three decades, the prison population grew from 196,000 inmates in 1972[32] to 2,033,331 today, making the United States, the land of the free and home of the brave, the world's largest jailer.[33] Even before the mass incarceration frenzy began in the mid-1970s, prisoners were predominantly black and Latino, many of whom were political activists who organized resistance movements in their communities. Currently, as a result of the war on drugs, one in three young black males is likely to spend some time in the criminal justice system, in spite of the fact that drug use is relatively the same across racial and ethnic groups. In fact, the United States is:

> incarcerating African-American men at a rate approximately four times the rate of incarceration of black men in South Africa under apartheid. Worse still, we have managed to replicate—at least on a statistical level—the shame of chattel slavery in this country: The number of black men in prison . . . has already equaled the number of men enslaved in 1820. . . . [And] if current trends continue, only fifteen years remain before the United States incarcerates as many African-American men as were forced into chattel bondage at slavery's peak, in 1860.[34]

The race to incarcerate in turn broke up families (where parental rights weren't dissolved altogether), increased poverty and unemployment, and denied ex-felons the right to public housing, food stamps, and veterans' benefits—leading to more crime and more poverty in poor communities. The net result was to push young black men a little closer to prison, the madhouse, or the grave.

That objective conditions worsened for the majority of black Americans in both the post-Reconstruction and post–civil rights eras seems difficult to contest in the face of such overwhelming evidence. Yet, the rationalizations for such immiseration, for right-wing politicians and pundits, offer a unique opportunity to condemn impoverished communities of color rather than Draconian public policies. To add insult to injury, Du Bois notes that the alleged faults and failures of southern Reconstruction were placed squarely on "Negro ignorance and corruption."[35] Citing literally dozens of accounts from children's history texts, Du Bois discovers an overwhelming chorus of agreement on this issue. "The South found it necessary to pass Black Codes," writes one historian,

> for the control of the shiftless and sometimes vicious freedmen. The Freedman's Bureau caused the Negroes to look to the North rather than the South for support and by giving them a false sense of equality did more harm than good. With the scalawags, the ignorant and non-property holding Negroes under the leadership of the carpetbaggers, engaged in a wild orgy of spending in the legislature. The humiliation and distress of the Southern whites was in part relieved by the Ku Klux Klan, a secret organization which frightened the superstitious blacks.[36]

As a result of such organized ideological assault, Du Bois writes, "There is scarce a child in the street that cannot tell you that the whole effort was a hideous mistake ... that the history of the U.S. from 1866 to 1876 is something of which the nation ought to be ashamed and which did more to retard and set back the American Negro than anything that has happened to him."[37]

The political history of the 1960s is, similarly, currently subject to a great deal of revision. Tragically, the debates aren't merely academic, as these selective narratives have sanctioned drastic changes in, if not the shredding of, the social contract. One of the primary architects of the ideological campaign to roll back the welfare state was Charles Murray, who in his 1984 book *Losing Ground* recast the advances of civil rights in the 1960s as a veritable bargain with the devil:

> The reforms of the 60s ... discouraged poor young people, and especially poor young males, from pursuing this slow, incremental approach [to lifting oneself out of poverty] in four ways. First, they increased the size of the welfare package and transformed the eligibility rules so as to make welfare a more available and attractive *temporary* alternative to a job. Second, the reforms in law enforcement and criminal justice increased access to income from the underground economy. By the 1970s, illegal income (including that from dealing in drugs, gambling, and stolen goods, as well as direct predatory

crime) had become a major source of income in poor communities. Third, the breakdown in inner-city education reduced job readiness. Acculturation to the demands of the workplace—arriving every day on time, staying there, accepting the role of a subordinate—diminished as these behaviors were no longer required in the schoolroom. Fourth, the reforms diminished the stigma associated with welfare and simultaneously devalued the status associated with working at a menial, low-paying job—indeed holding onto a menial job became in some communities a *source* of stigma.[38]

Thus Murray, the right-wing rhetorical alchemist, is able to reassign the blame of growing inner-city joblessness and the rise of underground economies from deindustrialization to glam welfare lifestyles. Similarly, failing schools do not indict the failure to enforce the *Brown* decision and decreasing federal financial support for education, but lead to unsubstantiated claims about the disappearance of discipline from urban schools. What Murray and other conservative pundits achieved was a successful transformation of social problems into narratives of individual moral failing and pathology. Indeed, the political impact of such challenges to the legacy of the 1960s on a largely white electorate, as we have seen, was pure gold.

As the ongoing assaults against black communities suggest, the transition from the political economy of slavery to that of industrialism and postindustrialism does not necessarily translate into a progressive movement from slavery to freedom, but rather signals a shift in racial definition and management, from brute force to the rule of law.[39] Given the increasing intolerance for overt physical violence, we should not be surprised by the relentless focus on legal structures in periods of racial reaction. As in the post-emancipation era, the post–civil rights government responded to the challenge of equal, popular participation and power-sharing with the shoring up of a new form of power: a reactionary Supreme Court was put in place to achieve what neither a divided Congress nor the presidency could. With the appointment of Morrison R. Waite as chief justice under Ulysses S. Grant the legal protections guaranteed by the Fourteenth and Fifteenth Amendments were both "reinterpreted" and rendered innocuous. Racialized exclusion was not

made explicit, but rather implicitly upheld in the all-too-familiar
rhetoric of "states' rights." These and other mechanisms of systemic, violent exclusion remained in place until the 1960s when
black Americans fought for and won—again—the rights and
entitlements they had garnered as citizens 100 years prior. With the
appointment of Antonin Scalia and the promotion of William
Rehnquist to chief justice of the Supreme Court under Reagan—
and the subsequent addition of Clarence Thomas under George
Bush, Sr.—the civil rights legislation ensuring equal opportunity in
education and work, and enfranchisement have been substantially
undermined. Though the discourse of "states rights" is alive and
well, in the rush to repudiate explicit forms of racist oppression
once legally sanctioned by *Plessy v. Ferguson*, the courts inaugurated
a new commitment to state racelessness—or colorblindness—as a
means to camouflage the "post-racist racism" of the state while
aiding the simultaneous advancement of market exclusions in the
rapidly expanding private sector.[40] Nowhere is the influence of this
decidedly conservative court more keenly felt than around educational equality and access. To the degree that struggles for democratization require an educated and empowered citizenry, the right-wing
attack on public education at all levels makes strategic sense.

Black Educational Exclusion in the Post–Civil Rights Era

Access to and influence upon educational institutions were central
to the revolutionary efforts of those short years (post-1863 and
post-1964) when black political power was a visible reality.
According to Du Bois, newly manumitted slaves desired only two
things: first they wanted land to own and work for their own crops,
and second they wanted to know, not just "cabalistic letters and
numbers" but also the "meaning of the world," and more specifically "what . . . had recently happened to them—this upturning
of the universe and revolution of the whole social fabric."[41]
Consumed with a desire for learning, black Americans poured
themselves into organizing and introduced free public schooling to
the South where none had existed before—one of many astounding
results of the political will and social vision of former slaves turned

critical and active citizens, not even a generation removed from bondage. Yet in the decades that followed, efforts to undermine black civil rights and entitlements would become equally organized, culminating in the 1896 *Plessy* decision asserting the constitutionality of already "separate but [un]equal" transportation, school facilities, and the like. As a result of this landmark decision, blacks would not be denied educational access altogether as in the days of slavery, but, as Du Bois notes, there were innumerable ways to make such schools run considerably less efficiently:

> ... in the first place, the public school funds were distributed with open and unashamed discrimination. Anywhere from twice to ten times as much was spent on the white child as was on the Negro child, and even then the poor white child did not receive an adequate education. ... The Negro schools were given few buildings and little equipment. No effort was made to compel Negro children to go to school. On the contrary, in the country they were deliberately kept out of school by the requirements of contract labor which embraced the labor of the wife and children as well as of the laborer himself. The course of study was limited. The school term was made and kept short and in many cases there was the deliberate effort, as expressed by one leading Southerner, Hoke Smith, when two Negro teachers applied for a school, to "take the less competent."[42]

The significance of the attack on educational opportunity was underscored by Du Bois, who noted the preeminent role of education even in the midst of counterrevolution: "Had it not been for the Negro school and college," he wrote, "the Negro would, to all intents and purposes, have been driven back to slavery."[43]

In the 1960s, black Americans agitated for and achieved not only the formal desegregation of the nation's public school system but also expanded access to higher education for all students of color, and gained influence over the curriculum, including black faculty appointments. Currently, conservatives have enacted an even broader range of strategies to weaken public schools, and as a result black educational access at all levels is in jeopardy. To be sure, older tactics used to mitigate the potential for children to learn in school are still in play—the wide disparities in school funding, the squalid and dangerous conditions of most school buildings, the overcrowding, the lack of adequate resources, and the hiring of non-certified and unmotivated teachers that Du Bois details. But with the dawn of the twenty-first century, school reforms have made the experience of schooling, now compulsory, as painful as possible, challenging even the most invested learners. Such measures include the militarization of schools—now complete with security guards, drug-sniffing dogs, see-through knapsacks, metal detectors, and zero-tolerance policies that threaten those who misbehave not only with expulsion but actual jail time. Added to this are the hijacking of the curriculum in the name of test preparation, the culturally biased nature of such examinations, and the "accountability" measures under the No Child Left Behind act, which pressure school administrators to get rid of those students who test poorly and might threaten the school's survival. Such pressure played an important role in the Houston School System, held up as a model by President George W. Bush, which not only did nothing to prevent students from leaving school but also falsified dropout data in order for principals to get financial bonuses and meet district demands. Tamar Lewin and Jennifer Medina report in *The New York Times* that large numbers of students who are struggling academically are being pushed out of New York City schools in order to not "tarnish the schools' statistics by failing to graduate on time."[44] The relentless instrumentalization of knowledge in the interests of testing and accountability has proved a venerable means of short-circuiting debates over the very substance of school curricula and the place of the student within it. As a result of such reform efforts, prospects for poor and minority youth to attend higher education are rapidly worsening.

Once a central cause of the civil rights movement, the rolling back of educational access for black students has been met with near total silence, both in the mainstream media and in the academy. Gary Orfield, head of the Harvard Project on School Desegregation, notes that:

> During the civil rights movement, research on desegregation was abundant. Government and foundations pumped dollars into race relations work. It seemed as if the academic world was a strong

resource for the ... movement, but it turned out to be only a fair-weather friend. When the government was supporting civil rights, the issue became the central focus of research. Once politics changed and research funding dried up, so did most academic involvement. Part of the logic of resegregation is the cutoff of most of the information about segregation and its consequences. The federal government has published no basic statistics on national school segregation levels since the Carter administration.[45]

And as the university goes, so it seems goes public discussion. In the post–civil rights era, local administrators and school boards, of course, never say that they are pursuing a separate-but-equal educational system. Rather, they discuss the need to move beyond "physical desegregation" or "racial balancing" or "numerical integration." In spite of such blatant Orwellian mystifications, challenges to the resegregation of public schools are conspicuously absent in mainstream media. In fact, on educational policy issues, with the exception of affirmative action, the Democratic and Republican party leaderships enjoy a convergence of opinion and purpose, with support for standards, accountability, and school choice registering near universal approval.

Although progressives see a partial victory in the Supreme Court's June 2003 affirmative action decision, it remains a shameful compromise that will provide little advantage to poor minorities. Nor should the decision be abstracted from decades of school-related decisions quietly denying black educational access. The fact remains that higher education remains out of reach for the vast majority of poor youth, who are subject to grossly inferior and rapidly resegregating elementary and secondary schools—schools that they are compelled by law to attend. All of which is to say that there might be no cause for a debate over alleged racial preferences now or in 25 years, as Justice O'Connor fancies, if the Rehnquist court had upheld and enforced the *Brown v. Board of Education of Topeka* verdict of 1954. But in a series of decisions since *Millikin v. Bradley* in 1974, the court, abetted by a largely silent academy, has quietly reversed the decree to desegregate established by the Warren court. With a team of graduate students, Gary Orfield has documented Supreme Court decision after decision—*Milliken v. Bradley II, Board of*

Education of Oklahoma v. Dowell (1988), *Freeman v. Pitts* (1992), and *Missouri v. Jenkins* (1995)—that have enabled the resegregation of the nation's schools such that they now resemble those of the *Plessy* era. In *Dismantling Desegregation*, Orfield observes:

> The common wisdom passed down by teachers through the genera-
> tions is that *Brown v. Board of Education* corrected an ugly flaw in
> American education and American law. We celebrate *Brown* and
> Martin Luther King Jr. in our schools, even when these very schools
> are almost totally segregated by race and poverty. Millions of
> African American and Latino students learn the lessons of Brown
> while they sit in segregated schools in collapsing cities, where almost
> no students successfully prepare for college.[46]

Over a decade ago, Jonathan Kozol attempted to awaken the conscience of the nation to this tragic denial of King's dream in *Savage Inequalities: Children in America's Schools*. Documenting in lurid detail the crushing inequalities between rich (predominantly white) and poor (predominantly nonwhite) school districts across the nation, Kozol exploded conventional wisdom that characterized public schools as the "great levelers" of a democratic society. According to this mythology, schools provide the conditions for hardworking youth graced with a little native intelligence to achieve the much-vaunted American dream. Yet the presumption of equality is entirely misguided:

> A typical wealthy suburb in which homes are often worth more than
> $400,000 draws upon a larger tax base in proportion to its student
> population than a city occupied by thousands of poor people.
> Typically, in the United States, very poor communities place high pri-
> ority on education, and they often tax themselves at higher rates than
> do the very affluent communities. But even if they tax themselves at
> several times the rate of an extremely wealthy district, they are likely
> to end up with far less money for each child in their schools.[47]

The consequence of such "savage inequality" is that poor school districts have had to forego (as they currently do) experienced, qualified teachers and up-to-date textbooks—let alone technologies like

VCRs or computers—and quite often a safe and healthy school infrastructure. Yet all public school children have to take the same standardized tests to gain access to a post-secondary educational credential, now an essential ingredient (though hardly a guarantee) for transcending minimum-wage work. According to the General Accounting Office, it would take $112 billion to bring the nations' public schools simply up to building code. That figure does not include monies for hiring certified, competent teachers, administrators, and support staff, or providing school children with adequate resources like books and computers or "extras" like courses in art and music, busing, and extracurricular sports. Tragically, the state of public schools has only deteriorated further since Kozol's publication in 1992, in keeping with the general decline in public support for matters of racial justice. Gary Orfield notes that "Among whites, though support for desegregation continued, the issue of racial justice went to the bottom of the list of national priorities. In 1995, 56 percent of whites thought that blacks were well off or better off than whites in terms of education in spite of massive gaps."[48] Explaining the contradiction between perception and the actual state of black education, Orfield points to the rhetoric of race: "Conservative politicians won white voters by telling them that civil rights policies had gone too far and were hurting whites. No powerful defense of civil rights and no leadership helping the public understand the persisting inequality in educational opportunities for minority students existed."[49]

Part of the reason for this ongoing crisis in American public schooling lies in federal cuts in education since the Reagan administration. The rationale for such a shift in national priorities is that American public schools are bureaucratic, wasteful, and altogether ineffectual—the result of a "big government" monopoly on education. As a result of such inefficiency, the public school system according to the famed Reagan-era study *A Nation at Risk*, poses a threat to national security and U.S. economic dominance in the world market. To be sure, some public schools are really ailing, but the reasons for this, according to Berliner and Biddle, authors of *The Manufactured Crisis*, have to do with the grossly unequal funding of public education, residential segregation, the astonishingly high poverty rates of U.S. school children relative to most other

industrialized nations, and inadequate health care and social services. Preferring the former diagnosis of general ineptitude, the
current administration insists that throwing money at schools will
not cure public school ills and will no longer be tolerated.

Rather than address the complexity of educational inequalities
disproportionately impacting poor and minority students, the Bush
administration sought solutions to troubled public schools in the
much touted No Child Left Behind (NCLB) legislation, which
afforded certain key advantages to constituencies in favor of privatization and resegregation, all the while appearing sympathetic to
the plight of poor and minority youth. Not only do they maintain
the advantages accorded white students, who perform better on
average than black and Latino students on standardized tests; the
proposed school reforms were also very business friendly. Renamed
No Child Left Untested by critics, the reform places high priority
on accountability, tying what little federal monies schools receive
to improved test performance. For additional financial support,
public schools are left no other meaningful option than engaging in
public/private partnerships, like the highly publicized deals cut
with soft drink giants, which provide schools with needed revenue
in exchange for soda machines in cafeterias. Similarly, media giants
who own the major publishing houses will benefit from the
52 million-strong market of public school students now required to
take tests every year from the third grade on. The impact of NCLB
also proved highly televisable, visibility now being a key factor in
the art of persuading a public weaned from political debate in favor
of spectacle. Thus the media provides routine reportage of school
districts' grade cards, public—often monetary—rewards given to
those schools that score high marks on achievement tests, and
liquidation of those that don't. Media preoccupation with school
safety issues, moreover, ensures highly publicized expulsion and
sometimes felony incarceration of troublemakers, typically
students of color. In short, accountability for teachers and administrators and zero tolerance for students who commit even the most
minor infractions are the new educational imperatives. All of which
demonstrates that the federal government is "doing something" to
assuage public fears about the problems of the nation's schools,
which are largely created through financial deprivation and

government policies favoring resegregation. As a result, the little federal aid schools do receive is increasingly spent on testing and prep materials as well as new safety measures, such as metal detectors, armed guards, security cameras, and fencing, in accordance with NCLB. In addition to draining public schools financially, both high-stakes testing and zero-tolerance policies have served to push out or kick out black and Latino youth in disproportionate numbers, as has been extensively documented by Henry Giroux in *The Abandoned Generation*, William Ayers et al. in *Zero Tolerance*, and Gary Orfield and Mindy Kornhaber in *Raising Standards or Raising Barriers?*[50]

Most recently it has become evident, as Du Bois would have predicted, that *all* children eventually suffer from the systemic disinvestment in education and other public goods and services at the hands of a right-wing, pro-business, and anti–civil rights governing elite. For example, the disastrous state of California's economy can be traced back, in part, to draconian cuts in education since the 1970s. The key factor in rising California spending in recent years—the alleged reason for the state's budgetary woes—has been its efforts to rebuild a crumbling educational system. Economist Paul Krugman explains that the passage of Proposition 13 in 1978, which introduced a cap on property taxes, "led to a progressive starvation of California's once-lauded public schools. By 1994, the state had the largest class sizes in the nation; its reading scores on par with Mississippi's."[51] So it seems the chickens came home to roost, as the infamous tax revolt of the 1970s fueled by racist propaganda dressed up as fiscal populism utterly devastated the state. According to Mike Davis, the famed author of *City of Quartz* and chronicler of Los Angeles's savage history,

> As the Latino population soared, white voters—egged on by rightwing demagogues—withdrew support from the public sector. California became a bad school state in lockstep with becoming a low wage state. Overcrowded classrooms and dangerous playgrounds are part of a vicious cycle with sweatshops and slum housing.[52]

As Californians sought to halt creeping "Mississippization," they passed, in addition to living-wage ordinances and other legislation, Proposition 98, which allocated more money for schools. This is what conservatives are now deriding as "runaway government spending." What began as a mechanism to perpetuate racialized exclusions in the post–civil rights era has led, in part, to the dramatic decline of the world's fifth-largest economy.

America's youth are not only paying the price of racial animus and political demagoguery at home, but also for the U.S. imperialist agendas abroad, particularly the military occupation of Iraq. Budgetary shortfalls in most states, exacerbated by the cost of the Iraqi war and security measures after 9/11, have only widened inequalities in funding, resulting in the mass firing of teachers, shortened school years, the dismantling of extracurricular programs, and the postponement of much-needed structural repair. Soaring deficits and the request of an additional $87 billion from Congress in October 2003 to aid the "peace" in Iraq signaled even more trouble for the nation's schools in the months ahead. Senator Robert Byrd reminded the president of his commitment to America's most vulnerable children in the following terms:

> It is equally ironic that the Administration is seeking an estimated $60 to $70 billion in additional funding for Iraq from the American taxpayers at a time when the Senate is debating adding a fraction of that amount to an appropriations bill to provide critical funding— funding the President himself pledged to provide in his No Child Left Behind initiative—for schoolchildren in poor school districts.[53]

Ironic indeed, as little mention has been made of repealing the president's most recent round of tax cuts, primarily for the wealthiest 1 percent of the population, to offset military expenditures.

The tragic state of public education in America is not unrelated to the future of higher education. Clearly, children's K-12 experiences play a determining role in their access to and preparedness for post-secondary education. Recently, the academy has come under fire for low retention rates among minority youth—more a pretense for another round of cuts in federal funding and student aid than drawing public attention to a serious concern—yet few critics seem willing to acknowledge the obvious. Higher education is successful only to the degree that K-12 education is successful.[54] Poor and minority youth who manage to survive the deplorable conditions of their K-12 education and still want to continue their schooling face skyrocketing college tuition rates, which have more than doubled in the last decade. Moreover, the government recently revised the formula for financial aid for colleges, which will reduce the nation's largest primary award program, the Pell grant, by $270 million once it takes effect in the 2004–2005 academic year.[55] For youth unable to afford the costs, the Supreme Court 2003 decision to uphold affirmative action is rather meaningless, as relatively few apply to the 100 or so most selective colleges in the United States where such policies are in effect. A recent study by Anthony Carnevale, vice president of the Educational Testing Service, found that "74 percent of the students at the 146 most prestigious colleges and universities—where competition for admissions is most intense and where affirmative action is practiced—come from the top 25 percent of the nation's socioeconomic scale (as measured by income, educational attainment, and occupation of parents). Only 3 percent come from the bottom 25 percent, and a total of 10 percent come from the bottom half."[56] Given such expenses, the same tiering in K-12 is visible at the post-secondary level, where a credential from a typical Ivy League university like Brown (at $38,000 a year) will open doors for its graduates in the Fortune 500 in the same way that a credential from a local community college will likely qualify one to join the ranks of overworked and underpaid laborers in the service sector. Because college admissions officers tend to rely on hard variables like testing, and race and socioeconomic status are both more strongly correlated with high test scores than intelligence or aptitude, the distribution of scarce slots at highly selective universities is skewed in favor of white youth whose parents have money. Hence, Carnevale asserts that higher education, especially at public institutions that are supported largely through tax dollars, has become "a gift the poor give to the rich."[57]

According to a July 2003 article in the *Chronicle of Higher Education*, President Bush is planning to use the renewal of the Higher Education Act as an occasion to lambast universities for high tuition and dropout rates. As Bush looks to revive his "compassionate conservative" image in the upcoming election year with an issue that will play well with the public, political observers note that he can do that "by empathizing with low- and middle-income families that are struggling to pay their college bills. He can also do that, they say, by scolding colleges for allowing so many disadvantaged students to drop out lacking the skills they need to improve their lives."[58] The White House is right to be concerned, but tuition has gone up largely in response to successive cuts in the federal budget that the president himself signed into law. The anticipated report, "The College Cost Crisis," appeared in mid-September 2003. Written by John Boehner and Howard McKeon, two Republicans on the House education committee, the report charges the university with "wasteful spending," the result in part of a woeful lack of accountability "to parents, students, and taxpayers—the consumers of higher education."[59] The answer, the Congressmen believe, is in a bill that will further cut federal financing to colleges whose tuition hikes are more than double the rate of inflation or the consumer price index. Rather than meeting the needs of struggling students, the bill is simply a means to withdraw more funds from universities already so financially strapped they have had to compromise, as a matter of survival, the quality of education students receive by closing departments, offering fewer courses, hiring more grad students and adjuncts to teach courses, skimping on advising, health, and counseling services, and disbanding sports teams. Stanley Fish, Dean of the College of Liberal Arts and Sciences at the University of Illinois at Chicago, didn't miss the irony. It is precisely because of diminished federal support for education that colleges and universities are becoming cost-prohibitive for the working- and middle-class families that the government seems so eager to help. As applications for admissions continue to rise, financial support from the government has been withdrawn. Fish concludes: "If the revenues sustaining your operation are sharply cut and you are prevented by law from raising prices, your only recourse is to offer an inferior product. Those who say, as the state has said to the University of Illinois, 'We're taking $200 million from you but we expect you to do the job you were doing and do it even better,' are trafficking in either fantasy or hypocrisy."[60]

But there is one further irony. At the state level, monies are increasingly tied up in efforts to maintain military operations abroad as well as in a shift in priorities from education to incarceration domestically, particularly for blacks and Latinos. If the president really wanted to aid struggling poor and minority youth in their efforts to achieve a post-secondary degree, he might consider a repeal of the Drug-Free Student Aid provision of the Higher Education Act of 1998, a recent line of attack in the infamous War on Drugs. Under this ruling, any student who has been convicted of the possession or sale of a controlled substance is either temporarily—or perhaps permanently, depending on the offense— ineligible for Stafford loans, Pell grants, or work-study programs. Students with one drug possession conviction lose their aid for a year from the date of conviction; with two convictions, they lose two years; and a third offense results in permanent loss of aid. Sanctions for selling drugs are even stricter. The inherent unfairness of the law has been well documented by critics. It primarily impacts minority students of lower income, who are disproportionately tar- geted in the War on Drugs and are, unlike their middle-class coun- terparts, dependent on federal aid for schooling. Further, the law ignores any financial aid applicants who have committed crimes unrelated to drugs. For example, students found convicted of bombing a nursery school, shooting an eighth-grade teacher, or committing rape or armed robbery remain eligible for student loans; yet those who have been caught smoking a joint or two are refused, their life's ambition reduced to enticing customers to super-size their orders of fries. Failing to differentiate between degrees of drug use and abuse, or between victims and villains, the message this conveys about national priorities is chilling. Graham Boyd concludes: "The government is creating two classes of peo- ple: One class to whom we want to give an education and succeed in life, and another class of low-income drug users who we want to relegate to a life of working at McDonalds."[61]

What the Drug-Free Student Aid provision makes clear is the government's obvious preference for incarcerating black youth over educating them. It's well-documented that drug war enforcement is racist. African Americans make up only 12 percent of the U.S. population and only 13 percent of drug offenders—proportionately about the same as white drug users. Yet, African Americans make up 62 percent of those with drug convictions, prompting Boyd's observation that "Since . . . the provision is more about who gets convicted by the system than it is about the drug offense, it's much more likely that you'll lose your funding if you are black than white."[62] Like the Black Codes of post-Reconstruction, punishment for drug violation is not well correlated with crime, but rather race and class. The 2000–2001 school year was the first in which the drug-free provision of the Higher Education Act was enforced (ex-inhaler-in-chief Bill Clinton didn't require question 35—which asks about any drug convictions, felony, or misdemeanor, state or federal—on the student aid application to be answered). As a result, about 34,000 students and college applicants were denied financial aid, mostly preventing black students from exercising one of the most basic principles of empowerment in the country.

Again, it is not simply black and brown youth who pay for the drug war and mass incarceration, but all youth—the eventual inheritors of a $5 trillion deficit and an utterly divided, unequal society. In fact, the recent Supreme Court decision upholding affirmative action at the University of Michigan Law School should not overshadow the court's "other" affirmative action, decision regarding the legality of "three strikes laws" in the criminal justice system (the only place, famed activist Angela Davis once quipped, with a robust affirmative action scheme). With well over 2 million inmates in the system whose sentences are now getting longer, and with the costs of maintaining a single prisoner at about $26,000 a year (triple that if they're older, and age they will), incarceration costs taxpayers about the same as an Iraqi war brought home—particularly when one adds on prison construction costs, medical costs, families reduced to welfare and children in foster care, and the loss of tax revenue at all levels. And that doesn't begin to gauge the destruction of poor and minority communities hardest hit by the prison boom. According to Marc Mauer, director of the Sentencing Project and author of *Race to Incarcerate*, there is a direct correlation between increases in state appropriations for criminal justice and decreases in spending on welfare, health care, and education—especially higher education.

Racial Pedagogy: Resurrecting the Language of Political Democracy

The reversal of democratic fortunes described by Du Bois in the penultimate chapter of *Black Reconstruction* entitled "Back toward Slavery" seems as relevant today as it did 70 years ago. He wrote:

> The attempt to make black men American citizens was in a certain sense all a failure, but a splendid failure. It did not fail where it was expected to fail. It was *Athanasius contra mundum*, with back to the wall, outnumbered ten to one, with all the wealth and all the opportunity, and all the world against him. And only in his hands and heart the consciousness of a great and just cause; fighting the battle of all the oppressed and despised humanity of every race and color, against the massed hirelings of Religion, Science, Education, Law and brute force.[63]

He further lamented the utter lack of organized progressive response, noting, "there is scarcely a bishop in Christendom, a priest in the church, a president, a governor, mayor, or legislature in the United States, a college professor or public school teacher, who does not in the end stand by War and Ignorance as the main method for the settlement of our pressing human problems. And this despite the fact that they deny it with their mouths every day."[64] The same silence on issues of racial equality and racial justice both nationally and internationally dominates contemporary mainstream political culture. Yet the university remains a crucial site of struggle and one of the few remaining spaces where a generation of young people can learn to assume the responsibilities of democratic citizenship. By way of conclusion, we would like to engage, as we have throughout this book, the challenges confronting intellectuals who attempt to foster a critical racial politics on campus, in spite of the university's much celebrated and much maligned "multicultural turn."

If the rollback of black educational access at all levels of schooling has been met with thunderous silence on the part of academics over the last two decades, so too have most reneged on their responsibility to engage students politically,[65] to begin an ethical dialogue rooted in a form of historical recovery that, in Du Bois's words, transcends "history for our pleasure and amusement, for inflating our national ego, and giving us a false but pleasurable sense of accomplishment."[66] This is not to suggest that racial politics were utterly avoided at a time in university history derided by conservatives as the "Great Takeover" of radical-tenured anti-Americans. But it is to acknowledge that beginning in the 1970s, critical analyses of race migrated from the social sciences to the humanities, specifically English departments—the primary site of the culture wars of the past two decades. The "multicultural turn" in fields like literary studies offered, at least initially, a radical reformulation of the experiences of blacks, Latinos, and other racialized minorities in the United States, but the project reflects only a partial victory at best. When the history of African Americans becomes literary history, the privatization of racial experience reproduces, instead of challenges, the neoliberal emphasis on hyperindividualism and its depoliticizing effects. The upshot of such a relentless focus on identity politics over and against an incisive historical analysis of U.S. efforts to contain and control black populations remains, nonetheless, in keeping with the conservative ideology of colorblindness and its commitment to historical denial. By reconceptualizing racism as a private—as opposed to deeply political and structural—phenomenon, colorblind ideology displaces the tensions of contemporary racially charged relations to the relative invisibility of the private sphere—safely beyond the reach of public policy intervention. More to the point, it remains well beyond the reach of most students (apart from a few business majors eager to tap new minority markets) who were left alone to ponder why such privatized experiences, so removed from their own, should concern them. The consequences of efforts to "manage diversity" were not only that a generation or two of students were without any sense of how race structures U.S. society both currently and historically;[67] it also played a central role in a deepening disdain for "big government," "welfare," and programs for "special interests," resulting in a political sensibility that begins and ends with how to keep the tax man out of one's pockets.

Riding the wave of P.C. backlash, students on campuses across the country (many, like Penn State, remain less than 4 percent minority) have repeatedly challenged, and even organized against, diversity requirements, politically correct curricula, and teachers who assume that "politicized" notions of race, class, and gender are proper analytical categories for engaging texts. Why does race have to be "injected" into discussions of American literature or popular culture? Why do we have to pay special attention to black or Latino/a writers rather than focusing on whoever happens to be the best author? Shouldn't the curriculum—and the classroom—be colorblind? It seldom takes long for the focus of such questions to shift to students' ambivalence about public life and politics outside the classroom, becoming: Why does race have to be "injected" into national politics, when the focus should be the common good and things that unify us? Why are minority groups given special treatment and special preference rather than being held accountable to an ethic of "individual initiative" and "personal responsibility"? Of course, students' attempts to find social and political relevance in the cultural texts they study are entirely commendable. University educators often suggest confidently, if a little vaguely, that the subjects we teach will help students figure out or act on their world in some way. Engaging cultural texts is relevant, we insist, and then leave it to them to figure out how. And while the students' interventions are clearly reactionary to ears more sensitive to the nuances of political argument, their questions reflect at some profound level what it means for young people to be engaged pedagogically by the mainstream national political scene of the post–civil rights era.

The movement from textual analysis to the contexts of students' everyday lives is necessary if one believes that knowledge should be related to broader public discourses in order to create the conditions for students to bridge the gap between what they learn in the university and how they become critical social agents. Civic education in this sense is rooted in a pedagogical commitment to making knowledge both relevant and meaningful through rigorous critical engagement, while at the same time suggesting that the production of knowledge cannot be divorced from civic responsibility and social action. Yet such a pedagogy is particularly fraught with difficulties when it involves taking up combustible issues such as race, for teachers no less than for students. Teachers, according to the prevailing common sense, have to be "professional," which in this instance means nonpolitical, objective, neutral, and disinterested. Yet it should be clear that neutrality is already a kind of advocacy for the status quo. For students, mastering close reading and technical proficiency present one kind of challenge; critically engaging the experiences, beliefs, knowledge, and affective investments with which they come to class is quite another.

Student investment in how to read a novel by William Faulkner or even a film by Spike Lee, however, is not at all the same as their investment in the contemporary civic mythologies designed to seduce and flatter them. We are not simply referring to the white-washed, Disney-like images of a golden past filled with valiant, freedom-loving heroes that persistently invade the collective psyche. We mean as well those narratives of national identity—of American dreams achieved through rugged individualism and a solid work ethic—that have been in place for decades, but whose meanings have radically shifted since the Reagan era. Standing on what they see as a firm and principled commitment to individual initiative and motivation, personal responsibility, and race-neutral institutions and values, many students feel deeply suspicious of traditional liberals and leftists who recklessly "inject race" into political debate, defend "race-based" social programs, and organize constituencies and movements "by race."[68] In this way, the current neoliberal commitment to hyperindividualism functions symbolically as a challenge to the privileges and "handouts" allegedly bestowed on minority groups. In other words, "individualism" serves as an antonym for "race," yet reinvents a racial logic that serves the interests of white populations at the same time. Without question, students' responses to discussions of race indicate that "The political public sphere and the electorate have indeed been contoured according to 'race' and racial identity. But the constituency whose beliefs and fears have been most significantly molded to their racial identity [since] the 1980s are *whites*."[69]

In attempting to connect what goes on in the classroom to what is happening on the national scene, the real pedagogical challenge is, first, how to get students to recognize that racism is not simply a function of private discrimination. Rather, racism continues to play a dramatic role in contemporary public dialogues about culture, politics, and citizenship, in spite of the dominant perception that they now live in a colorblind society. Second is the related challenge of exploring how race is used in national debates as one strategy to fuel popular antistatist and antipolitical sentiments, which have opened the door to increasing privitization, depoliticization, and racial segregation of the body politic. Such rhetoric has enabled not only widespread racial backlash, but also the nearly unchecked ascendancy of neoliberal corporate power, the widening of the gap between the rich and the poor, the most regressive tax reform in the nation's history, and the wholesale dismantling of the welfare state at a time when increasing numbers of citizens—particularly young citizens—need it more than ever. Current debates over who has rights under what conditions and who doesn't, who has the capacity to be productive and self-governing and who hasn't, who can meaningfully participate in public life and who can't, have been recoded in overtly racist terms that students and the general public alike often resist recognizing as such. We hear, for example, of the need to support victims' rights over criminals' rights, to cut taxpayers a break over tax recipients, to transform welfare into back-to-workfare programs, and to end affirmative action in support of "race-transcending" public policies. In each instance, reform is tacitly understood to improve the conditions of a white electorate at the expense of people of color, while racist assumptions that equate criminality and "dependancy" with blackness remain unchallenged.

To be sure, the declining social and economic standing of many lower- and middle-income whites and blacks since the 1980s is real and a not-unanticipated result of the ascendancy of neoliberal social and economic policies and the racist backlash that quickened the dismantling of public goods and services as it criminalized the poor. Over the last two decades, conservative rhetoric has consistently invoked the evils of state power (particularly the imagined abuses against white taxpayers), the need to transfer decision-making from the federal to the local level to allegedly "rejuvenate" the public sphere, and the need to dismantle massive bureaucracy. In the post-9/11 era, the Right demands patriotism, insisting that educators (the "weak link" in the War on Terror) who focus critically on the nation's segregationist and imperialist history are anti-American terrorist sympathizers.[70] Republicans thus continue to hammer home the virtues of the "free market," accountability, the return of family values, and rugged individualism. At the same time, they work assiduously to curb the participation of citizens in public life and "to strengthen some of the most authoritarian and oppressive features of the state (the military, police, prison system, control over personal life)"—a point even more salient in the era of permanent war and expanding American empire.[71] Thus conservatives have created, in the cruelest of ironies, popular support for policies that merely deepen racialized exclusions, economic disequilibrium, social dislocation, and generalized anxieties and fears among the citizenry. In short, the ongoing attack on "big government" has aided and abetted a dramatic restructuring of the corporate economy that took every advantage of decentralization, deregulation, and privatization, while furthering the impoverishment and despair of minority populations. As Carl Boggs points out, "the real thrust behind appeals for smaller government is to severely weaken the social and popular side of the state and to legitimate an assault on 'welfare,' popular constituencies (women and minorities), and 'special interests' (labor) that might impede the global developmental objectives of corporate planners."[72] Ironically, he notes, as conservatives rail against "big government," they simultaneously seek to "broaden state control over potentially insurgent groups" through increased spending on the military, police, prison building, and mechanisms for surveillance.[73]

The result of such depoliticization is deep popular despair and cynicism about even the possibility of political action—feelings that often translate into "ever more privatized lifestyles . . . [and] deep hostility toward the public sphere in general."[74] Such sentiments seem particularly evident among youth. A January 2001 report in the *Chronicle of Higher Education* suggested that "political engagement among first-year students has reached an all-time low. . . . Only 28.1 percent of entering college students reported an interest in 'keeping up to date with political affairs,' the lowest level since the survey was established in 1966, when the figure was 60.3 percent."[75] Among the reasons for this atrophy of political interest, the report locates at least three. First, a shift in youth's priorities, as "nearly three-quarters of first-year students indicated that they want to be 'very well off.'" Second, a tendency not just to tune out of national politics, but to "look inward," suggesting "students are focused much more locally and even individually on their own circumstances." And finally a sense, according to one student, that many youth feel alienated from politics, stating "The issues don't affect them. . . . A lot of people just think that politics nowadays [is] disgusting."[76] The legacy of the last 20 years for this generation of youth, then, is not only an eventual confrontation with fearsome economic uncertainty, an increasing gap between the rich and poor, the dismantling of the welfare state, and racial resegregation, but also a deepening crisis in public life and the practice of democracy itself.

The task facing critical educators is not an easy one. The university itself is under attack—facing pressure by the corporate sector to instrumentalize knowledge in the interests of profit and pressure by conservatives to cleanse humanities curricula of any critique of American culture or political institutions, particularly those that challenge the much vaunted "racial harmony" of the post–civil rights era or the desire of the government to bring democracy (as opposed to a form of neocolonial occupation) to the Middle East. In addition, access to a post-secondary school as well as social and cultural capital for growing numbers of working-class and minority students is clearly becoming more questionable. Linking questions of pedagogy to political agency requires that educators mediate the troubled relations between knowledge and action, private concerns and public interests, and individual freedoms and the social contract. At the same time, any pedagogical project that seeks to revitalize questions of citizenship, community, and the public good must not only be attentive to the ways in which such notions have historically perpetuated racist exclusions, but also the degree to which they are currently in danger of disappearing in an era marked by the ascendancy of neoliberalism, which Bourdieu once referred to as the "logic of pure market."

By addressing the contemporary crisis of democracy through a rigorous historical and social analysis of the contradictory relation between democratic government and the market economy, as well as to the class, gender, and racial divisions of society (particularly in light of the official rhetoric of "colorblindness" that shapes much public policy), we can open up a space for imagining a more just democracy. Specifically, our task as educators is to open up dialogue by resurrecting the public memory of racial oppression and exclusion. This suggests using public memory as a way to bear witness to human suffering and to challenge the addiction to racial amnesia that has become a hallmark of the present era. In doing so, we can "assess the public morality of American social policy" as we engage students in the exploration of possible, more democratic arrangements for government, the economy, and civil society, as well as those changes in consciousness, culture, and education needed to sustain such reforms.[77] In doing so, we can arrest the rhetorical transformation of the public sphere by once again invoking, after W. E. B. Du Bois, a language of critical historical inquiry, substantive democracy, and racial justice both at home and across the globe. In this way we can hope to reverse the desperate experience of fear, anxiety, uncertainty, and alienation that accompany the painful erosion of individual and social agency. In challenging the atrophy of a public discourse of racial equality, racial justice, and substantive democracy, it is possible to challenge the disappearance of politics itself.

Du Bois's curious blend of genealogy and prophecy in *Black Reconstruction* should remind us that the promise of political democracy can only be achieved by a sustained pedagogical engagement with the nation's most cherished values of freedom, justice, and equality, *situated in and challenged by* its history of racial apartheid and class exploitation. Through sustained historical analysis and ethical inquiry, we can begin to understand the racist policies and practices of the past as they continue to shape our present. But there is more to this legacy than the history of racism; there are also the hard-won struggles of those who opposed racial exploitation and exclusion, and it is this aspect of public memory that must also be engaged and acted upon. That is Du Bois's message, and hopefully our legacy for future generations.

Part III

Incorporating Education and Shredding the Social Contract

Chapter 6

Youth, Higher Education, and the Breaking of the Social Contract: Toward the Possibility of a Democratic Future

Children are the future of any society. If you want to know the future of a society look at the eyes of the children. If you want to maim the future of any society, you simply maim the children. The struggle for the survival of our children is the struggle for the survival of our future. The quantity and quality of that survival is the measurement of the development of our society.[1]

—Ngugi Wa Thiong'o

Youth and the Crisis of the Future

Any discourse about the future has to begin with the issue of youth because more than any other group, youth embody the dreams, desires, and commitment of a society's obligations to the future. This echoes a classical principle of liberal democracy, in which youth both symbolized society's responsibility to the future and offered a measure of its progress. For most of the twentieth century, Americans have embraced as a defining feature of politics that all levels of government would assume a large measure of responsibility for providing the resources, social provisions, security, and modes of education that simultaneously offered young people a future as well as expanded the meaning and depth of a substantive democracy. Youth not only registered symbolically the importance of modernity's claim to progress, they also affirmed the importance of the liberal, democratic tradition of the social contract, in which adult responsibility was mediated through a willingness to fight for the rights of children, enact reforms that invested in their future, and provide the educational conditions necessary for them to make use of the freedoms they have while learning how to be critical citizens, all the while enabling the reproduction of that society. Within such a project, democracy was linked to the well-being of youth, while the status of how a society imagined democracy and its future was contingent on how it viewed its responsibility toward future generations.

But the category of youth did more than affirm modernity's social contract rooted in a conception of the future in which adult commitment and intergenerational solidarity were articulated as a vital public service; it also affirmed those vocabularies, values, and social relations central to a politics capable of defending vital institutions as a public good and contributing to the quality of public life. Such a vocabulary was particularly important for higher education, whose highest ideals reflected the recognition that how it educated youth was connected to both the democratic future it hoped for and its claim as an important public sphere.

Yet, at the dawn of the new millennium, it is not at all clear that we believe any longer in youth, the future, or the social contract (even in its minimalist version). Since the Reagan/Thatcher revolution of the 1980s, we have been told that there is no such thing as society and, indeed, ever since that nefarious pronouncement, institutions committed to public welfare have been disappearing. Those of us who, against the conventional wisdom, insist on the relationship between higher education and the future of democracy, have to face a disturbing reversal in priorities with regard to youth and education under the reign of neoliberalism.[2] Rather than being cherished as a symbol of the future, youth are now seen as a threat to be feared and a problem to be contained. A seismic change has taken place in which youth are now being framed as both a generation of suspects and a danger to public life. As we mentioned in chapter 2, if youth once symbolized the moral necessity to address a range of social and economic ills, they are now largely portrayed as the source of most of society's problems. Hence, youth now constitute a crisis that has less to do with improving the future than with denying it. A concern for children is the defining absence in almost any discourse about the future, and the obligations this implies for adult society. To witness the abdication of adult responsibility to children we need look no further than the current state of children in America, who once served as a "kind of symbolic guarantee that America still had a future, that it still believed in a future, and that it was crucial to America to invest its faith in that future."[3]

No longer "viewed as a privileged sign and embodiment of the future,"[4] youth are now demonized by the popular media and derided by politicians looking for quick-fix solutions to crime, joblessness, and poverty. In a society deeply troubled by their presence, youth prompt a public rhetoric of fear, control, and surveillance, which translates into social policies that shrink democratic public spheres, highjack civic culture, and militarize public space. Equipped with police and drug-sniffing dogs, though not necessarily adequate teachers or textbooks, public schools increasingly resemble prisons. Students begin to look more like criminal suspects to be searched, tested, and observed, under the watchful eye of administrators who appear to be less concerned with educating them than with containing their every move. Nurture, trust, and respect now give way to fear, disdain, and suspicion. In many suburban malls, young people, especially urban youth of color, cannot shop or walk around without having appropriate identification cards or being in the company of a parent. Children have fewer rights than almost any other group and fewer institutions protecting these rights. Consequently, their voices and desires are almost completely absent from the debates, policies, and legislative practices that supposedly address their needs.

Instead of providing a decent education to poor young people, American society offers them the growing potential of being incarcerated, buttressed by the fact that the United States is one of the few countries in the world that sentences minors to death and spends "three times more on each incarcerated citizen than on each public school pupil."[5] Instead of guaranteeing our youth food, decent health care, and shelter, we serve them more standardized tests; instead of providing them with vibrant public spheres, we offer them a commercialized culture in which consumerism is the only obligation of citizenship. But in the hard currency of human suffering, children pay a heavy price. In one of the richest democracies in the world, 20 percent of children are poor during the first three years of life and more than 12.2 million live in poverty; 9.2 million children lack health insurance; millions lack affordable child care and decent early childhood education; in many states more money is being spent on prison construction than on education; and the infant mortality rate in the United States is the highest of any industrialized nation.[6] When broken down along racial categories, the figures become even more despairing. For example, "In 1998, 36 percent of black and 34 percent of Hispanic children lived in poverty, compared with 14 percent of white children."[7] In some cities, such as the District of Columbia, the child poverty rate is as high as 45 percent.[8] While the United States ranks first in military technology, military exports, defense expenditures, and the number of millionaires and billionaires, it is ranked eighteenth among the advanced industrial nations in the gap between rich and poor children, twelfth in the percent of children in poverty, seventeenth in the efforts to lift children out of poverty, and twenty-third in infant mortality.[9] One of the most shameful figures on youth as reported by Jennifer Egan, a writer for The New York Times, is that "1.4 million children are homeless in America for a time in any given year ... and these children make up 40 percent of the nation's homeless population."[10] In short, economically, politically, and culturally, the situation of youth in the United States is intolerable and obscene. It is all the more unforgivable since President Bush insisted during the 2000 campaign that "the biggest percentage of our budget should go to children's education." He then pushed for a series of budgets in which 40 times more money went for tax cuts for the wealthiest 1 percent of the population than to education.[11] But Bush's insensitivity to American children represents more than a paean to the rich, since he also signed a punitive welfare reform bill that requires poor, young mothers to work a 40-hour week—while at the same time cutting low-income childcare programs. It gets worse.

While the Bush administration aims to spend up to $400 billion on defense, not including the $4 billion per month needed to cover the costs of postwar occupation and construction in Iraq, it allocates only $16 billion to welfare.[12] At the same time that Congress has passed tax cuts amounting to $723 billion, 50 percent of which will go to the richest 1 percent of the population, it also slashed $14.6 billion in benefits for veterans, $93 billion in Medicaid cuts, and promoted cuts in student loans, education programs, school lunches, food stamps, and cash assistance for the elderly, poor, and disabled.[13]

Youth have become the main target onto which class and racial anxieties are projected. Their very presence in a neoliberal age where there is "no such thing as society" represents both the broken promises of democracy and the violation of a social contract that traditionally offered young people the right to decent food, education, health, employment, and other crucial rights fundamental to their survival, dignity, and a decent future. Corporate deregulation and downsizing and a collective fear of the consequences wrought by systemic class inequalities, racism, and a culture of "infectious greed" have created a generation of displaced and unskilled youth who have been expelled from the "universe of moral obligations."[14] Youth within the economic, political, and cultural geography of neoliberal capitalism occupy a degraded borderland in which the spectacle of commodification exists side by side with the imposing threat of the prison-industrial complex and the elimination of basic civil liberties. As neoliberal policies dissociate economics from social costs, "the political state has become the corporate state."[15] Under such circumstances, the state does not disappear, but, as Pierre Bourdieu has brilliantly reminded us,[16] is refigured as its role in providing social provisions, intervening on behalf of public welfare, and regulating corporate plunder is weakened. The neoliberal state no longer invests in solving social problems; it now punishes those who are caught in the downward spiral of its economic policies. Punishment, incarceration, and surveillance represent the face of the new state. Social guarantees for youth as well as civic obligations no longer represent an important priority in the public imagination. Similarly, as market values supplant civic values, it becomes increasingly difficult "to translate private worries into public issues and, conversely, to discern public issues in private troubles."[17] Alcoholism, homelessness, poverty, and illiteracy, among other issues, are not seen as social but as individual problems—matters of character, individual fortitude, and personal responsibility. As we have stressed throughout this book, in light of the increased antiterrorism campaign waged by the Bush administration, it becomes easier to militarize domestic space, criminalize social problems, and escape from the responsibilities of the present while destroying all possibilities of a truly democratic future. The social costs of the complex cultural and economic effects of this assault can no longer be ignored by educators, parents, and other concerned citizens.

The war against youth, in part, can be understood in terms of the practices of a rapacious, neoliberal capitalism. For many people today, the private sphere has become the only space in which to imagine any sense of hope, pleasure, or possibility. Culture as an activity in which people actually produce the conditions of their own agency through dialogue, community participation, resistance, and political struggle is being replaced by a "climate of cultural and linguistic privatization"[18] in which culture becomes something you consume and the only kind of speech that is acceptable is that of the savvy shopper. Neoliberalism, with its emphasis on market forces and profit margins, narrows the legitimacy of the public sphere by redressing social concerns through privatization, deregulation, consumption, and safety. Ardent consumers and disengaged citizens provide fodder for a growing cynicism and depoliticization of public life at a time when there is an increasing awareness of corporate corruption, financial mismanagement, and systemic greed, as well as the recognition that a democracy of critical citizens is being replaced quickly by an ersatz democracy of consumers. The desire to protect market freedoms and wage a war against terrorism at home and abroad, ironically, has not only ushered in a culture of fear but has also dealt a lethal blow to civil freedoms. Resting in the balance of this contradiction is both the fate of democracy and the civic health and future of generations of children and young adults.

In this insufferable climate of increased militarization, repression, and unrestrained exploitation, young people become the new casualties in an ongoing war against justice, freedom, citizenship, and democracy. Lawrence Grossberg argues that "the current rejection of childhood as the core of our social identity, is at the same time, a rejection of the future as an affective investment."[19] But the crisis of youth not only signals a dangerous state of affairs for the future, it also portends a crisis in the very idea of the political and ethical constitution of society and the possibility of articulating the relevance of democracy itself. It is in reference to this crisis that we want to address the relationship between higher education and the future.

Higher Education and the Crisis of the Social

There is a distinguished tradition of educational thought in the United States extending from Thomas Jefferson and W. E. B. Du Bois to Jane Addams, John Dewey, and C. Wright Mills, in which the future of the university is premised on the recognition that in order for freedom to flourish in the public realm, citizens have to be educated for the task of self-government. Jane Addams and John Dewey, for example, argued that public and higher education should provide the conditions for people to involve themselves in the most pressing problems of society, to acquire the knowledge, skills, and ethical responsibility necessary for "reasoned participation in democratically organized publics."[20] C. Wright Mills challenged schooling as a form of corporate training and called for fashioning higher education within a public philosophy committed to a radical conception of citizenship, civic engagement, and public wisdom.[21] Education in this context was linked to public life through democratic values such as equality, justice, and freedom, rather than as an adjunct of the corporation, whose knowledge and values were defined largely through the prism of commercial interests. Education was crucial to individual agency and public citizenship, and integral to defending the relationship between an autonomous society—rooted in an ever-expanding process of self-examination, critique, and reform—and autonomous individuals, for whom critical inquiry is propelled by the ongoing need to pursue ethics and justice as matters of social conscience and public good. In many ways, the academy has remained faithful, at least in theory, to a project of modern politics whose purpose was to create citizens capable of defining and implementing universal goals such as freedom, equality, and justice as part of a broader attempt to deepen the relationship between an expanded notion of the social and the enabling ground of a vibrant democracy.

In the last two decades, a widespread pessimism about public life and politics has developed in the United States. Individual interests now outweigh collective concerns and market ideals have taken precedence over democratic values. Moreover, the ethos of citizenship has been stripped of its political dimensions and is now reduced to the obligations of consumerism. In the vocabulary of neoliberalism, the public collapses into the personal, and the personal becomes "the only politics there is, the only politics with a tangible referent or emotional valence,"[22] and it is within such an utterly personalized discourse that human actions are shaped and agency is privatized. Under neoliberalism, hope disappears or is diminished as the public sphere atrophies and, as Peter Beilharz argues, "politics becomes banal, for there is not only an absence of citizenship but a striking absence of agency."[23] As power is increasingly separated from traditional politics and public obligations, corporations are less subject to the control of the state and "there is a strong impulse to displace political sovereignty with the sovereignty of the market, as if the latter has a mind and morality of its own."[24] Under the auspices of neoliberalism, the language of the social is either devalued or ignored altogether as the idea of the public sphere is equated with a predatory space, rife with danger and disease—as with "public" restrooms, "public" transportation, and urban "public" schools, as we argued in chapter 2. Tellingly, the term *public* has itself become pejorative; it is little wonder that the remaining functions of the state are organized around the military and the police. Dreams of the future are now modeled on the narcissistic, privatized, and self-indulgent needs of consumer culture and the dictates of the allegedly free market. Mark Taylor, a social critic turned apologist for the market, both embodies and captures the sentiment well with his comment: "Insofar as you want to engage practice responsibly, you have to play with the hand you're dealt. And the hand we're dealt seems to me to be one in which the market has certainly won out over other kinds of systems."[25] There is more at stake here than another dominant media story about a left academic who finally sees the entrepreneurial light. The narrative points to something much larger. Samuel Weber has suggested that what seems to be involved in this transformation is "a fundamental and political redefinition of the social value of public services in general, and of universities and education in particular."[26]

Within this impoverished sense of politics and public life, the university is gradually being transformed into a training ground for the corporate workforce, rendering obsolete any notion of higher education as a crucial public sphere in which critical citizens and democratic agents are formed. As universities become increasingly strapped for money, corporations provide the needed resources for research and funds for endowed chairs, getting in exchange a powerful influence on both the hiring of faculty and how research is conducted and for what purposes. In addition, universities now offer up buildings and stadiums as billboards for brand-name corporations in order to procure additional sources of revenue while also adopting the values, management styles, cost-cutting procedures, and the language of "excellence" that has been the hallmark of corporate culture. The boundaries between commercial culture and public culture have become blurred as universities rush to embrace the logic of industrial management while simultaneously forfeiting those broader values central to a democracy and capable of limiting the excesses of corporate power. Although the university has always had ties to industry, there is a new intimacy between higher education and corporate culture, characterized by what Larry Hanley calls a "new, quickened symbiosis."[27] As Masao Miyoshi points out, the result is "not a fundamental or abrupt change perhaps, but still an unmistakable radical reduction of its public and critical role."[28]

How do we understand the university in light of both the crisis of youth and the related crisis of the social under neoliberalism? How can the future be conceptualized given the erosion of the social and public life over the last 20 years and the corporatization of higher education? Any concern about the future of the university has to both engage and challenge this transformation, reclaiming the role of the university as a democratic public sphere. In what follows, we want to analyze the university as a corporate entity within the context of a crisis of the social. In particular, we will focus on how this crisis is played out not only through the erosion of public space, but through the less explained issues of public versus corporate time, on the one hand, and the related issues of agency, pedagogy, and public mission on the other.

Public Time Versus Corporate Time

Questions of time are crucial to how a university structures its public mission, influencing the role of faculty, the use of space, student access, and the organization of particular forms of knowledge, research, and pedagogy. Time is not simply a category for understanding the future, but is also used to legitimate particular social relations and make claims on human behavior—representing one of the most important battlefields for determining how the future of higher education is played out in political and ethical terms. As a theoretical construct in addition to a lived reality, time refers not only to the way in which temporality is mediated differently by institutions, administrators, faculty, and students, but also how it shapes and allocates power, identities, and space through a particular set of codes and interests. More important, time is a central feature of politics, and orders not only the pace of the economy, but also the time available for consideration, contemplation, and critical thinking. When reduced to a commodity, time often becomes the enemy of deliberation and thoughtfulness and undermines the ability of political culture to function critically.

For the past 20 years, time as a value and the value of time have been redefined through the dictates of neoliberal economics, which has largely undermined any notion of public time guided by the noncommodified values central to a political and social democracy. As Peter Beilharz observes, "time has become our enemy. The active society demands of us that we keep moving, keep consuming, experience everything, travel, work as good tourists more than act as good citizens, work, shop, and die. To keep moving is the only way left in our cultural repertoire to push away . . . meaning. . . . [and consequently] the prospects and forms of social solidarity available to us shrink before our eyes."[29]

Without question, the future of the university will largely rest on the outcome of the current struggle between the university as a public space with the capacity to slow time down in order to question what Jacques Derrida calls the powers that limit "a democracy to come,"[30] and a corporate university culture wedded to a notion of accelerated time in which the principle of self-interest replaces politics and consumerism replaces a broader notion of social agency. A meaningful and inclusive democracy is indebted to a notion of public time, while neoliberalism celebrates what we call corporate time. In what follows, we want to briefly comment on some of the theoretical and political work suggested by each of these notions of time and the implications they have for addressing the future of higher education. Public time as a condition and critical referent makes visible how politics is played out through the unequal access different groups have to "institutions, goods, services, resources, and power and knowledge."[31] That is, it offers a critical category for understanding how the ideological and institutional mechanisms of higher education work to grant time to some faculty and students and to withhold it from others; how time is mediated differently within different disciplines and among diverse faculty and students; how time can work across the canvas of power and space to create new identities and social formations capable of "intervening in public debate for the purpose of affecting positive change in the overall position and location in society."[32] When linked to issues of power, identity, ideology, and politics, public time can be an important social construct for orienting the university toward a vision of the future in which critical learning becomes central to increasing the scope of human rights, individual freedom, and the operations of a substantive democracy. In this instance, public time resonates with a project of leadership, teaching, and learning in which higher education seems an important site for investing democratic public life with substance and vibrancy.

Public time rejects the fever-pitch appeals of "just in time" or "speed time," demands often made within the context of "ever faster technological transformation and exchange,"[33] and reflecting corporate capital's golden rule: "time is money." Public time slows time down, not as a simple refusal of technological change or a rejection of all calls for efficiency but as an attempt to create the institutional and ideological conditions that promote long-term analysis, historical reflection, and deliberation over what our collective actions might mean for shaping the future. Rejecting an instrumentality that evacuates questions of history, ethics, and justice, public time fosters dialogue, thoughtfulness, and critical exchange. Public time offers room for knowledge that contributes to society's self-understanding, that enables it to question itself, and seeks to legitimate intellectual practices that are not only collective and noninstrumental but deepen democratic values while encouraging pedagogical relations that question the future in terms that are political, ethical, and social. As Cornelius Castoriadis points out, public time puts into question established institutions and dominant authority, rejecting any notion of the social that either eliminates the question of judgment or conceals the question of responsibility.[34] Rather than maintaining a passive attitude toward power, public time demands and encourages forms of political agency based on a passion for self-governing actions informed by critical judgment and a commitment to linking social responsibility and social transformation. Public time legitimates those pedagogical practices that provide the basis for a culture of questioning and social engagement, a culture that offers students the knowledge, skills, and social practices necessary for resistance, a space of translation, and a proliferation of discourses. Public time unsettles common sense and disturbs authority while encouraging critical and responsible leadership. As Roger Simon observes, public time "presents the question of the social— not as a space for the articulation of pre-formed visions through which to mobilize action, but as the movement in which the very question of the possibility of democracy becomes the frame within which a necessary radical learning (and questioning) is enabled."[35] Put differently, public time affirms a politics without guarantees and a notion of the social that is open and contingent. Public time also provides a conception of democracy that is never complete and determinate but constantly open to different understandings of the contingency of decisions, mechanisms of exclusions, and operations of power.[36] Public time challenges neoliberalism's willingness to separate the economic from the social, politics from power. It also challenges neoliberalism's failure to address human needs and social costs.

At its best, public time renders governmental power explicit, and in doing so it rejects the language of secrecy, absolutes, and the abrogation of the conditions necessary for the assumption of basic freedoms and rights. Moreover, public time considers civic education the basis, if not essential dimension, of justice because it provides individuals with the skills, knowledge, and passions to talk back to power, while simultaneously emphasizing both the necessity to question that accompanies political agency and the assumption of public responsibility through active participation in the process of governing. Expressions of public time in higher education can be found in shared notions of governance between faculty and administration, in forms of collegiality tied to vibrant communities of exchange and the furthering of democratic values, and in pedagogical relations in which students do not just learn about democracy but experience it through a sense of active participation, critical engagement, and social responsibility. The notion of public time has a long history in higher education and has played a formative role in shaping some of the most important principles of academic life. Public time, in this instance, registers the importance of pedagogical practices that provide the conditions for a culture of questioning in which teachers and students engage in critical dialogue and unrestricted discussion in order to affirm their role as social agents, inspect their own past, and engage the consequences of their own actions in shaping the future.

As higher education becomes increasingly corporatized, public time is replaced by corporate time. In corporate time, the "market is viewed as a 'master design for all affairs,'"[37] profitmaking defines responsibility, and consumption is the privileged site for determining value between the self and the larger social order. Corporate time fosters a narrow sense of leadership, agency, and public values, and is largely indifferent to those concerns that are critical to a just society but are not commercial in nature. The values of hierarchy, materialism, competition, and excessive individualism are enshrined under corporate time and play a defining role in how it allocates space, manages the production of particular forms of knowledge, guides research, and regulates pedagogical relations. Hence, it is not surprising that corporate time accentuates privatized and competitive modes of intellectual activity, largely removed from public obligations and social responsibilities.

Divested of any viable democratic notion of the social, corporate time measures relationships, productivity, space, and knowledge according to the dictates of cost efficiency, profit, and a market-based rationality. Time, within this framework, is accelerated rather than slowed down and reconfigures academic labor increasingly through (though not limited to) new computer technologies, which are making greater demands on faculty time, creating larger teaching loads, and producing bigger classes. Under corporate time, speed controls and organizes place, space, and communication as a matter of quantifiable calculation. And as Peter Euben observes, under such circumstances a particular form of rationality emerges as common sense:

> When speed rules so does efficient communication. Calculation and logic are in, moral imagination and reasoned emotions are out. With speed at a premium, shorthand, quantification and measurements become dominant modes of thought. Soon we will talk in cliches and call it common sense and wisdom.[38]

While we take up the corporatization of the university in more detail in chapter 7, we want to comment on some of the ways in which corporate time structures the culture of university life. Corporate time maps faculty relationships through self-promoting market agendas and narrow definitions of self-interest. Caught on the treadmill of getting more grants, teaching larger classes, and producing more revenue for the university, faculty become another casualty of a business ideology that attempts to "extract labor from campus workers at the lowest possible cost, one willing to sacrifice research independence and integrity for profit."[39] Under corporatization, time is accelerated and fragmented. Overworked and largely isolated, faculty are now rewarded for intellectual activities privileged as entrepreneurial, "measured largely in the capacity to transact and consume."[40] Faculty are asked to spend more time in larger classrooms while they are simultaneously expected to learn and use new instructional technologies such as Powerpoint, the Web, and various multimedia pedagogical tools. Faculty now interact with students not only in their classes and offices, but also in chat rooms and through e-mail.

Grounded in the culture of competitiveness and self-interest, corporate time reworks faculty loyalties. Faculty interaction is structured less around collective solidarities built upon practices that offer a particular relationship to public life than through corporate-imposed rituals of competition and production that conform to the "narrowly focused ideas of the university as a support to the economy."[41] For instance, many universities are now instituting post-tenure review as an alleged measure of faculty accountability and an efficient way to eliminate "deadwood" professors. As Ben Agger points out, what is "Especially pernicious is the fact that faculty are supposed to axe their own colleagues, thus pitting them against each other and destroying whatever remains of the fabric of academic community and mutuality."[42]

Corporate time also fragments time by redefining the role of faculty as a form of academic labor in which part-time labor is pitted against "academic work as full-time commitment and career."[43] Under such conditions, faculty solidarities are weakened more and more as corporate time demands cost-efficient measures by outsourcing instruction to part-time faculty who are underpaid, overworked, lack health benefits, and deprived of any power to shape the conditions under which they work. Powerlessness breeds resentment and anger among part-time faculty, and fear and insecurity among full-time faculty, who no longer believe that their tenure is secure. Hence, the divide between part- and full-time faculty is reproduced by the heavy hand of universities as they downsize and outsource under the rubric of fiscal responsibility and accountability, especially in the post 9-11 era. But more is reproduced than structural dislocations among faculty; there is also a large pool of crippling fear, insecurity, and resentment that makes it difficult for faculty to take risks, forge bonds of solidarity, engage in social criticism, and perform as public intellectuals rather than as technicians in the service of corporate largesse.

Leadership under the reign of corporate culture and corporate time has been reconceived as a form of homage to business models of governance. As Stanley Aronowitz points out, "Today . . . leaders of higher education wear the badge of corporate servants proudly."[44] Gone are the days when university presidents were hired for intellectual status and public roles. College presidents are now labeled as Chief Executive Officers, and are employed primarily because of their fundraising abilities. Deans of various colleges are often pulled from the ranks of the business world and pride themselves on the managerial logic and cost-cutting plans they adopt from the corporate culture of Microsoft, Disney, and IBM. Bill Gates, the CEO of Microsoft, and Michael Eisner, the CEO of Disney, replace John Dewey and Robert Hutchins as models of educational leadership. Rather than defend the public role of the university, academic freedom, and worthy social causes, the new corporate heroes of higher education now focus their time and energies on selling off university services to private contractors, forming partnerships with local corporations, searching for new patent and licensing agreements, and urging faculty to engage in research and grants that generate external funds. Under this model of leadership the university is being transformed from a place to think to a place to imagine stock options and profit windfalls.

Corporate time provides a new framing mechanism for faculty relations and modes of production and suggests a basic shift in the role of the intellectual. Academics now become less important as a resource to provide students with the knowledge and skills they need to engage the future as a condition of democratic possibilities. In the "new economy," they are entrepreneurs who view the future as an investment opportunity and research as a strategic career move rather than as a civic and collective effort to improve the public good. Increasingly, academics find themselves being de-skilled as they are pressured to teach more service-oriented and market-based courses and devote less time to their roles either as well-informed public intellectuals or as "cosmopolitan intellectuals situated in the public sphere."[45]

Corporate time not only transforms the university as a democratic public sphere into a space for training while defining faculty as entrepreneurs; it also views students as customers, potential workers, and as a source of revenue. As customers, students "are conceptualized in terms of their ability to pay. . . . and the more valued customers are those who can afford to pay more."[46] One consequence, as Gary Rhoades points out, is that student access to higher education is "now shaped less by considerations of social justice than of revenue potential."[47] Consequently, those students who are poor and under-served are increasingly denied access to the benefits of higher education. Of course, the real problem, as Cary Nelson observes, is not merely one of potential decline, but "long term and continuing failure to offer all citizens, especially minorities of class and color, equal educational opportunities,"[48] a failure that has been intensified under the corporate university. As a source of revenue, students are now subjected to higher fees and tuition costs, and are bombarded by brand-name corporations who either lease space on the university commons to advertise their goods or run any one of a number of student services, from the dining halls to the university bookstore. Almost every aspect of public space in higher education is now designed to attract students as consumers and shoppers, constantly subjecting them to forms of advertising mediated by the rhythms of corporate time, which keeps students moving through a marketplace of logos rather than ideas. Such hyper-commercialized spaces increasingly resemble malls, transforming all available university space into advertising billboards and bringing home the message that the most important identity available to students is that of the consuming subject. As the line between public and commercial space disappears, the gravitational pull of Taco Bell, McDonald's, Starbucks, Barnes and Noble, American Express, and Nike, among others, creates a "geography of nowhere,"[49] a consumer placelessness in which all barriers between a culture of critical ideas and branded products simply disappear.[50] Education is no longer merely a monetary exchange in which students buy an upscale, lucrative career, it is also an experience designed to evacuate any broader, more democratic notion of citizenship, the social, and the future that students may wish to imagine, struggle over, and enter. In corporate time, students are disenfranchised "as future citizens and reconstitute[d] . . . as no more than consumers and potential workers."[51]

Corporate time not only translates faculty as multinational operatives and students as sources of revenue and captive consumers; it also makes a claim on how knowledge is valued, how the classroom is organized, and how pedagogy is defined. Knowledge under corporate time is valued as a form of capital. As Michael Peters observes, entire disciplines and bodies of knowledge are now either valued or devalued on the basis of their "ability to attract global capital and . . . potential for serving transnational corporations. Knowledge is valued for its strict utility rather than as an end in itself or for its emancipatory effects."[52] Good value for students means taking courses labeled as "relevant" in market terms, which are often counterposed to courses in the social sciences, humanities, and the fine arts, which are concerned with forms of learning that do not readily translate into either private gain or commercial value. Under the rule of corporate time, the classroom is no longer a public sphere concerned with issues of justice, critical learning, or the knowledge and skills necessary for independent thought and civic engagement. As training replaces education, the classroom, along with pedagogy itself, is transformed as a result of the corporate restructuring of the university.

As the structure and content of education change, intellectual and pedagogical practices are less identified with providing the conditions for students to learn how to think critically, hold institutional authority accountable for its actions, and act in ways that further democratic ideals. Rather than providing the knowledge and skills for asserting the primacy of politics, social responsibility, and ethics as central to preparing students to participate in democracy, intellectual practice is subordinated to managerial, technological, and commercial considerations. Not only are classroom knowledge and intellectual practice bought and traded as marketable commodities, but they are also defined largely within what Zygmunt Bauman calls "the culture of consumer society, which is more about forgetting, [than] learning."[53] That means forgetting that knowledge can be emancipatory, that citizenship is not merely about being a consumer, and that the future cannot be sacrificed to ephemeral pleasures and values of the market. When education is reduced to training, the meaning of self-government is devalued and democracy is rendered meaningless.

It is essential to recognize in the rise of corporate time that while it acknowledges that higher education should play a crucial role in offering the narratives that frame society, it presupposes that faculty, in particular, will play a different role and assume a "different relation to the framing of cultural reality."[54] Many critics have pointed to the changing nature of governance and management structures in the university as a central force in redefining the relationship of the university to the larger society, but little has been said about how the changing direction of the university impacts the nature of academic activity and intellectual relations.[55] At one level, the changes give greater control of academic life to administrators and an emerging class of managerial professionals, but also privilege those intellectuals in technological sciences whose services are indispensable to corporate power. Not all forms of information reign equally as commodities in the new economy. Academic labor is now prized for how it fuses with capital, rather than how it contributes to what Geoff Sharp calls "society's self-understanding."[56] The changing institutional and social forms of the university reject the elitist and reclusive models of intellectual practice that traditionally have refused to bridge the gap between higher education and the larger social order, theory and practice, the academic and the public. Within the corporate university, transformation rather than contemplation is now a fundamental principle for judging and rewarding intellectual practice. Divorced from social justice or democratic possibilities, transformation is defined through a notion of society that entirely privileges the material interests of the market. Higher education's need for new sources of funding neatly dovetails with the inexhaustible need on the part of corporations for new products. Within this symbiotic relationship, knowledge is directly linked to its application in the market, resulting in a collapse of the distinction between knowledge and the commodity. Knowledge has become capital to invest in the market but has little to do with the power of self-definition, civic commitments, or ethical responsibilities that "require an engagement with the claims of others"[57] and with questions of justice. At the same time, the conditions for scholarly work are being transformed through technologies that eliminate face-to-face contact, speed up the labor process, and define social exchange in terms that are more competitive, instrumental, and impersonal.

Electronic, digital, and image-based technologies shape notions of the social in ways that were unimaginable a decade ago. Social exchanges can now proceed without the presence of "real" bodies. Contacts among faculty and between teachers and students are increasingly virtual, yet these practices profoundly change the nature of the social in instrumental, abstract, and commodified terms. As John Hinkson and Geoff Sharp have pointed out, these new intellectual practices and technological forms are redefining the nature of the social in higher education such that the free sharing of ideas and cooperativeness as democratic and supportive forms of collegiality seem to be disappearing.[58] This is not just an issue that can be taken up strictly as an assault on academic labor; it also raises fundamental questions about where those values that support democratic forms of solidarity, sharing, dialogue, and mutual understanding are to be found in university life. This is an especially important issue since such values serve as a "condition for the development of intellectual practices devoted to public service."[59] The ethic of public service that once received some support in higher education is being eliminated—and with it those intellectual relations, scholarly practices, and forms of collegiality that leave some room for addressing a less commodified and democratic notion of the social.

In opposition to corporate time, instrumentalized intellectual practices, and a denatured view of the social, we want to reassert the importance of the academy as a site of struggle and resistance. Central to such a challenge is the necessity to define intellectual practice "as part of an intricate web of morality, rigor and responsibility"[60] that enables academics to speak with conviction, enter the public sphere to address important social problems, and demonstrate alternative models for bridging the gap between higher education and society. This notion of intellectual practice refuses both the instrumentality and privileged isolation of the academy, while affirming a broader vision of learning that links knowledge to the power of self-definition and the capacities of administrators, academics, and students to expand the scope of democratic freedoms, particularly as they address the crisis of the social as part and parcel of the crisis of both youth and democracy itself. Implicit in this notion of social and intellectual practice is a view of academics as public intellectuals. Following Edward Said, we are referring to those academics engaged in intellectual practices that question power rather than merely consolidate it, enter into the public sphere in order to alleviate human suffering (occurring through the effects of poverty, racism, environmental abuse, unsafe working conditions, etc.), make the connections of power visible, and work individually and collectively to create the pedagogical and social conditions necessary for what the late Pierre Bourdieu has called "realist utopias."[61] We want to conclude by taking up how the role of both the university as a democratic public sphere and the function of academics as public intellectuals can be further enabled through what we call a politics of educated hope.

Toward a Politics of Educated Hope

If the rise of the corporate university is to be challenged, educators and others need to reclaim the meaning and purpose of higher education as an ethical and political response to the demise of democratic public life. They need to insist on the role of the university as a public sphere committed to deepening and expanding the possibilities of democracy. This approach suggests new models of leadership based on the understanding that the real purpose of higher education is more than to help people get a lucrative job. Beyond this ever-narrowing instrumental justification, there are more relevant goals, such as opening higher education up to all groups; creating a critical citizenry; providing specialized work skills for jobs that really require them; democratizing relations of governance among administrators, faculty, and students; and taking seriously the imperative to disseminate an intellectual and artistic culture. Higher education may be one of the few sites left in which students learn how to mediate critically between democratic values and the demands of corporate power, between identities founded on democratic principles and identities steeped in forms of competitive, atomistic individualism that celebrate self-interest, profitmaking, and greed. Toni Morrison is right in arguing that "If the university does not take seriously and rigorously its role as a guardian of wider civic freedoms, as interrogator of more and more complex ethical problems, as servant and preserver of deeper democratic practices, then some other regime or menage of regimes will do it for us, in spite of us, and without us."[62] Only if this struggle is taken seriously by educators and others can the university be reclaimed as a space of debate, discussion, and at times dissidence. Within such an educational space, time can be unconditionally apportioned to what Cornelius Castoriadis calls "an unlimited interrogation in all domains"[63] of society, especially with regard to the operations of dominant authority and power and the important issues that shape public life, enhancing practices that contribute to the ongoing process of democratization.

Higher education should be defended as a form of civic education in which teachers and students have the chance to resist and rewrite those modes of pedagogy, time, and rationality that refuse to include questions of judgment and issues of responsibility. This would include using teaching practices dictated by the state, working in overcrowded classrooms, teaching excessive numbers of classes, and being denied the requisite time to perform research. Understood this way, higher education is neither a consumer-driven product nor a form of training and career preparation but a mode of critical education that renders all individuals fit "to participate in power. . . . to the greatest extent possible, to participate in a common government,"[64] to be capable as Aristotle said of both governing and being governed. If higher education is to bring democratic public culture and critical pedagogy back to life, educators need to provide students with the knowledge and skills that enable them not only to judge and choose between different institutions, but also to create those institutions they deem necessary for living lives of decency and dignity. Education should provide not only the tools for citizen participation in public life, but also for exercising leadership. As Castoriadis insists, "People should have not just the typical right to participate; they should also be educated in every aspect (of leadership and politics) in order to be able to participate"[65] in governing society.

Reclaiming higher education as a public sphere begins with the crucial project of challenging corporate ideology and its attendant notion of time, which covers over the crisis of the social by dissociating all discussions about the goals of higher education from the realm of democracy. This project points to the important task of redefining higher education as a democratic public sphere not only to assert the importance of the social, but also to reconfigure it so that "economic interests cease to be the dominant factor in shaping attitudes"[66] about the social as a realm devoid of politics and democratic possibilities. Education is about issues of work and economics, questions of justice, social freedom, and the capacity for democratic agency and change. It is also about the related issues of power, exclusion, time, and citizenship, and how these categories are shaped within the broader spheres of culture, work, and economics. These are educational and political issues and should be addressed as part of a broader concern for renewing the struggle for social justice and democracy. Such a struggle demands, as the writer Arundhati Roy points out, that as intellectuals we ask ourselves some very "uncomfortable questions about our values and traditions, our vision for the future, our responsibilities as citizens, the legitimacy of our 'democratic institutions,' the role of the state, the police, the army, the judiciary, and the intellectual community."[67]

While it is crucial for educators and others to defend higher education as a public good, it is also important to recognize that the crisis of higher education cannot be understood outside of the overall restructuring of the social and civic life. The death of the social, the devaluing of political agency, the waning of noncommercial values, and the disappearance of noncommercialized public spaces have to be understood as part of a much broader attack on public entitlements such as healthcare, welfare, and social security, which are being turned over to market forces and privatized so that "economic transactions can subordinate and in many cases replace political democracy."[68]

Against the increasing corporatization of the university and the advance of global capitalism, educators need to resurrect a language of resistance and possibility, a language that embraces an insurgent utopianism while constantly being attentive to those forces that seek to turn such hope into a new slogan or punish and dismiss those who dare look beyond the horizon of the given. Hope as a form of insurgent utopianism is one of the preconditions for individual and social struggle, and the ongoing practice of critical education in a wide variety of sites—the attempt to make a difference by being able to imagine a different society and a readiness to act in other ways. Educated hope is utopian, as Ruth Levitas observes, in that it is understood "more broadly as the desire for a better way of living expressed in the description of a different kind of society that makes possible that alternative way of life."[69] Educated hope also demands a certain amount of courage on the part of intellectuals in that it demands that they articulate social possibilities, address injustice as part of a broader attempt to contest the workings of oppressive power, undermine various forms of domination, and fight for alternative ways to imagine the future. This is no small challenge at a time in American history when jingoistic patriotism is one of the few obligations of citizenship beyond consumption and dissent is viewed increasingly as the refuge for those who support terrorists.

Educated hope as a utopian longing becomes all the more urgent given the bleakness of the times, but also because it opens horizons of comparison by evoking not just different histories but different futures; at the same time, it substantiates the importance of openness and skepticism while problematizing certainty, or as Paul Ricoeur has suggested, hope is "a major resource as the weapon against closure."[70] As a form of utopian thinking, educated hope provides a theoretical service in that it pluralizes politics by generating dissent against the claims of a false harmony, and it provides an activating condition for promoting social transformation. Jacques Derrida has observed that if higher education is going to have a future that makes a difference in promoting democracy, it is crucial for educators to take up the "necessity to rethink the concepts of the possible and the impossible."[71] What Derrida is suggesting is that educated hope provides a vocabulary for challenging the presupposition that there are no alternatives to the existing social order, while simultaneously stressing the dynamic, still unfinished elements of a democracy to be realized.[72]

Educated hope as a form of insurgent utopianism—with its emphasis on seeing beyond the present and its belief in the power of social agency and change—accentuates the ways in which the political can become more pedagogical and the pedagogical more political. In the first instance, pedagogy merges politics and ethics with revitalized forms of civic education that provide the knowledge, skills, and experiences enabling individual freedom and social agency. Making the pedagogical more political demands that educators become more attentive to the ways in which institutional forces and cultural power are tangled up with everyday experience. It means understanding how higher education in the information age now interfaces with the larger culture, how it has become the most important site for framing public beliefs and authorizing specific relations between the self, the other, and the larger society that often shut down democratic visions. Any viable politics of educated hope must tap into individual experiences while at the same time linking individual responsibility with a progressive sense of social agency. Politics and pedagogy alike spring "from real situations and from what we can say and do in these situations."[73] As an empowering practice, educated hope translates into civic courage as a political and pedagogical practice that begins when one's life can no longer be taken for granted. In doing so, it makes concrete the possibility for transforming higher education into an ethical commitment and public event that confronts the flow of everyday experience and the weight of social suffering with the force of individual and collective resistance and the promise of an ongoing project of democratic social transformation.

Emphasizing politics as a pedagogical practice and performative act, educated hope accentuates the notion that politics is played out not only on the terrain of imagination and desire, but is also grounded in material relations of power and concrete social formations through which people live out their daily lives. Freedom and justice, in this instance, have to be mediated through the connection between civic education and political agency, which presupposes that the goal of educated hope is not to liberate the individual from the social—a central tenet of neoliberalism—but to take seriously the notion that the individual can only be liberated through the social. Educated hope, if it is to be meaningful, should provide a link, however transient, provisional, and contextual between vision and critique, on the one hand, and engagement and transformation on the other. But for such a notion of hope to be consequantial it has to be grounded in a vision and notion of pedagogy that has some hold on the present—a pedagogy that is attentive to those contexts that shape everyday life.

The limits of the utopian imagination are related, in part, to the failure of academics and intellectuals in a variety of public spheres not only to conceive of life beyond profit margins, but to imagine what pedagogical conditions might be necessary to bring into being forms of political agency that might expand individual rights, social provisions, and democratic freedoms. Against such failures, it is crucial for educators to address utopian longings as anticipatory rather than messianic, as temporal rather than merely spatial, forward-looking rather than backward-looking. Utopian thinking in this view is neither a blueprint for the future nor a form of social engineering, but a belief that different futures are possible. Utopian thinking rejects a politics of certainty and holds open matters of contingency, context, and indeterminancy as central to any notion of agency and the future. In this view, it is only through education that human beings can learn about the injustices of the present and the conditions necessary for them to "combine a gritty sense of limits with a lofty vision of possibility."[74] Educated hope poses the important challenge of how to reclaim social agency within a broader discourse of ethical advocacy while addressing those essential pedagogical and political elements necessary for envisioning alternatives to global neoliberalism and its attendant forms of corporate time and its attendant assault on public time and space.

Educated hope takes as a political and ethical necessity the need to address what modes of education are required for a democratic future and further requires that we ask questions such as: What pedagogical projects, resources, and practices can be put into place that would convey to students the vital importance of public time and its attendant culture of questioning as an essential step toward self-representation, agency, and a substantive democracy? How might public time, with its imperative to "take more time," compel respect rather than reverence, critique rather than silence, while challenging the narrow and commercial nature of corporate time? What kinds of social relations necessarily provide students with time for deliberation, as well as spaces for critical exchange in which they can critically engage those forms of power and authority that speak directly to them both within and outside of the academy? How might public time, with its unsettling refusal to be fixed or to collapse in the face of corporate time, be used to create pedagogical conditions that foster forms of self- and social critique as part of a broader project of fostering alternative desires and critical modes of thinking and democratic agents of change? How to deal with these issues is a major question for intellectuals in the academy today; their importance resides not just in how they might provide teachers and students with the tools to fight corporatization in higher education, but also how they address the need for fundamental institutional change in the ongoing struggle for freedom and justice in a revitalized democracy.

Far from innocent, pedagogical practices operate within institutional contexts that carry great power in determining what knowledge is of most worth, what it means for students to wield authority, and how such knowledge relates to a particular understanding of the self and its relationship to both others and the future. Connecting teaching as knowledge production to teaching as a form of self-production, pedagogy presupposes not only a political and ethical project that offers up a variety of human capacities, it also propagates diverse meanings of the social. Moreover, as an articulation of and intervention in the social, pedagogical practices always sanction particular versions of what knowledge is valuable, what it means to know something, how to be attentive to the operations of power, and how we might construct a sense of ourselves, others, and our physical environment. In the broadest sense, pedagogy is a principle feature of politics because it provides the capacities, knowledge, skills, and social relations through which individuals recognize themselves as social and political agents. As Roger Simon points out, "talk about pedagogy is simultaneously talk about the details of what students and others might do together and the cultural politics such practices support."[75]

While many critical educators and social theorists recognize that education, in general, and pedagogy, more specifically, cannot be separated from the dual crisis of representation and political agency, the primary emphasis in many of these approaches to critical pedagogy suggests that its foremost responsibility is to provide a space where the complexity of knowledge, culture, values, and social issues can be explored in open and critical dialogue through a vibrant culture of questioning. This position is echoed by Judith Butler, who argues, "For me there is more hope in the world when we can question what is taken for granted, especially about what it is to be human."[76] Zygmunt Bauman goes further, arguing that the resurrection of any viable notion of political and social agency is dependent upon a culture of questioning whose purpose, as he puts it, is to "keep the forever unexhausted and unfulfilled human potential open, fighting back all attempts to foreclose and pre-empt the further unraveling of human possibilities, prodding human society to go on questioning itself and preventing that questioning from ever stalling or being declared finished."[77]

Central to any viable notion of critical pedagogy is its willingness to take seriously those academic projects, intellectual practices, and social relations in which students have the basic right to raise, if not define questions, both within and outside of disciplinary boundaries. Such a pedagogy also must bear the responsibility of being self-conscious about those forces that sometimes prevent people from speaking openly and critically, whether they are part of a hidden curriculum of racism, class oppression, or gender discrimination, or part of those institutional and ideological mechanisms that silence students under the pretext of a claim to professionalism, objectivity, or unaccountable authority. Crucial here is the recognition that a pedagogical culture of questioning is not merely about the dynamics of communication but also about the effects of power and the mechanisms through which it either constrains, denies, or excludes particular forms of agency—preventing some individuals from speaking in specific ways, in particular spaces, under specific circumstances. Clearly such a pedagogy might include a questioning of such diverse issues as the corporatization of the educational context itself, the role of foreign policy, the purpose and meaning of the burgeoning prison-industrial complex, and the decline of the welfare state. Pedagogy makes visible the operations of power and authority as part of its processes of disruption and unsettlement—an attempt, as Larry Grossberg points out, "to win an already positioned, already invested individual or group to a different set of places, a different organization of the space of possibilities."[78]

At its best, critical pedagogy is self-reflective, and views its own practices and effects not as pregiven but as the outcome of previous struggles. Rather than defined as a technique, method, or "as a kind of physics which leaves its own history behind and never looks back,"[79] critical pedagogy is grounded in a sense of history, politics, and ethics that uses theory as a resource to respond to particular contexts, problems, and issues. We want to suggest that as educators we need to extend this approach to critical pedagogy beyond the project of simply providing students with critical knowledge and analytical tools. While this pedagogical approach rightly focuses on the primacy of dialogue, understanding, and critique, it does not adequately affirm the experience of the social and the obligations of responsibility and social transformation. Such a pedagogy attempts to open up for students important questions about power, knowledge, and what it might mean for students to critically engage the conditions under which life is presented to them, but it does not directly address what it would mean (whatever their respective fields of study), for them to work to overcome those social relations of oppression that make living unbearable for those youths and adults who are poor, hungry, unemployed, and refused even basic social services.

Our view is that pedagogy is inevitably political. However, educators such as Jeffrey C. Goldfarb have argued that education should primarily be used to engage students in "the great conversation," enable them to "pay attention to their critical faculties," and provoke informed discussion.[80] But Goldfarb also believes that education should be free from politics, providing students ultimately with the tools for civic discussion without the baggage of what he calls debilitating ideology. But by denying the relationship between politics and education, Goldfarb has no language for recognizing how pedagogy itself is shot through with issues of politics, power, and ideology. In opposition to Goldfarb, we believe that teaching and learning are profoundly political practices, as is evident in the most basic pedagogical and educational concerns, such as: How does one draw attention to the different ways in which knowledge, power, and experience are produced under specific conditions of learning? How are authority and power individually and institutionally distributed in both the university and the classroom? Who produces classroom knowledge and for whom? Who determines what knowledge is included or excluded? What is the agenda that informs the production and teaching of knowledge? What are the social and ideological horizons that determine student access to classrooms, and privilege particular forms of cultural capital—ways of talking, writing, acting, dressing, and embodying specific racial, gendered, and class histories? How does one determine how politics is connected to everyday questions of identity, beliefs, subjectivity, dreams, and desires? How does one acknowledge, mediate, or refuse dominant academic values, pressures, and social relations? Goldfarb confuses politics with indoctrination and in doing so has no way of critically analyzing how his own intellectual practices are implicated in relations of power that structure the very knowledge, values, and desires that mediate his relations to students and the outside world. Consequently, his willingness to separate education from matters of power and politics runs the risk of reproducing their worst effects.

Goldfarb wants to deny the symbiotic relationship between politics and education, but the real issue is to recognize how such a relationship might be used to produce pedagogical practices that condition but do not determine outcomes, that recognize that "the educator's task is to encourage human agency, not mold it in the manner of Pygmalion."[81] A critical education should enable students to question existing institutions as well as to view politics as "a labor aimed at transforming desirable institutions in a democratic direction."[82] But to acknowledge that critical pedagogy is directed and interventionist is not the same as turning it into a religious ritual. Critical approaches to pedagogy do not guarantee outcomes or impose a particular ideology, nor should they. But they should make a distinction between a rigorous ethical and scholarly approach to learning implicated in diverse relations of power and those forms of pedagogy that belie questions of ethics and responsibility, while allowing dialogue to degenerate into opinion and academic methods into unreflective and damaging ideological approaches to teaching. Rather than deny the relationship between education and politics, it seems far more crucial to engage it openly and critically so as to prevent pedagogical relations from degenerating into forms of abuse, terrorism, or contempt immune from self-reflection and analysis. What must be prevented at all costs are pedagogical practices that either silence or humiliate students, or that treat knowledge and authority in a slavish and unquestioning manner.

We want to return to a theme we addressed in chapter 3 and emphasize that a pedagogy that simply promotes a culture of questioning says nothing about what kind of future is or should be implied by how and what educators teach; nor does it address the necessity of recognizing the value of a future in which liberty, freedom, and justice play a constitutive role. While it is crucial for education to be attentive to those practices in which forms of social and political agency are denied, it is also imperative to create the conditions in which forms of agency are available for students to learn not only how to think critically but to act on what they believe. People need to be educated for democracy not only by expanding their capacities to think critically, but also by preparing them to assume public responsibility through active participation in the process of governing and engaging important social problems. This suggests connecting a pedagogy of understanding with pedagogical practices that are empowering and oppositional, practices that offer students the knowledge and skills needed to believe that a substantive democracy is not only possible but is worth taking responsibility for and struggling over. Feminist and postcolonial theorist Chandra Talpade Mohanty highlights this issue by arguing that pedagogy is not merely about matters of scholarship and what should be taught but also about issues of strategy, transformation, and practice. In this instance, a critical pedagogy should

> get students to think critically about their place in relation to the knowledge they gain and to transform their world view fundamentally by taking the politics of knowledge seriously. It is a pedagogy that attempts to link knowledge, social responsibility, and collective struggle. And it does so by emphasizing risks that education involves, the struggles for institutional change, and the strategies for challenging forms of domination and by creating more equitable and just public spheres within and outside of educational institutions.[83]

Any viable notion of critical pedagogy has to foreground issues not only of understanding but also social responsibility and address the implications the latter has for both affecting students and for influencing a democratic society. As Vaclav Havel has noted, "Democracy requires a certain type of citizen who feels responsible for something other than his own well feathered little corner; citizens who want to participate in society's affairs, who insist on it; citizens with backbones; citizens who hold their ideas about democracy at the deepest level, at the level that religion is held, where beliefs and identity are the same."[84]

Pedagogy plays a crucial role in nurturing this sense of responsibility and the process of learning how to translate critique and understanding into civic courage, enabling students to render what they view as a private matter into a concern for public life. Responsibility underscores the political nature of educational practices and suggests both a different future and the possibility of a revitalized politics. Responsibility challenges the "just-in-time" ethos of neoliberalism with an investment in long-term goals and commitments. Responsibility makes politics and agency possible; it does not rest at understanding, but recognizes the importance of students becoming accountable for others through their ideas, language, and actions. Being aware of the conditions that cause human suffering and the deep inequalities that generate dreadfully undemocratic and unethical contradictions for many people is not the same as resolving them. If pedagogy is to be linked to critical citizenship and public life, it needs to provide the conditions for students to learn in diverse ways how to take responsibility for moving society in the direction of a more realizable but never finalized or complete project of democracy. In this case, the burden of pedagogy is linked to the possibilities of understanding and acting. For educators such a pedagogy means engaging knowledge and theory as a resource to enhance the capacity for civic action and democratic change.

The future of higher education is inextricably connected to the future that we make available to the next generation of young people. Finding our way to a more human future means educating a new generation of scholars who not only defend higher education as a democratic public sphere, but who also see themselves as both scholars and citizen activists willing to connect their research, teaching, and service to broader democratic concerns over equality, justice, and an alternative vision of what the university might be and what society might become.

Chapter 7

Neoliberalism Goes to College: Higher Education in the New Economy

The single most important question for the future of America is how we treat our entrepreneurs.

—George Gilder[1]

A new form of domination is emerging in our times that breaks with the orthodox method of rule-by-engagement and uses deregulation as its major vehicle: "a mode of domination that is founded on the institution of insecurity—domination by the precariousness of existence."

—Zygmunt Bauman[2]

Neoliberalism and Corporate Culture

The ascendancy of neoliberalism and corporate culture in every aspect of American life not only consolidates economic power in the hands of the few; it also aggressively attempts to break the power of unions, decouple income from productivity, subordinate the needs of society to the market, reduce civic education to job training, and render public services and amenities an unconscionable luxury. But it does more. It thrives on a culture of cynicism, insecurity, and despair. Conscripts in a relentless campaign for personal responsibility, Americans are now convinced that they have little to hope for—or gain from—the government, nonprofit public organizations, democratic associations, public and higher education, or other nongovernmental social agencies. With few exceptions, the project of democratizing public institutions and goods has fallen into disrepute in the popular imagination as the logic of the market and increasing militarization of public life undermine the most basic social solidarities and blunt intellectual curiosity and conviction. The consequences include not only a state representative of a few elite, corporate interests, but also the transformation of a democratic republic into a national security state. Philosopher Susan Buck-Morss comments on this loss of democratic control:

> But there is another United States over which I have no control, because it is by definition not a democracy, not a republic. I am referring to the national security state that is called into existence with the sovereign pronouncement of a "state emergency" and that generates a wild zone of power, barbaric and violent, operating without democratic oversight, in order to combat an "enemy" that threatens the existence not merely and not mainly of its citizens, but of its sovereignty. The paradox is that this undemocratic state claims absolute power over the citizens of a free and democratic nation.[3]

The incessant calls for self-reliance and security that now dominate public discourse betray a weakened state that neither provides reasonable assurance that terrorist acts can be contained nor an adequate safety net for its populace, especially those who are young, poor, or marginalized. In short, private interests trump social needs, and economic growth becomes more important than social justice. The resulting shredding of the social contract is mediated through the force of corporate power and commercial values that dominates those competing public spheres and value systems that are critical to a just society and to democracy. The liberal democratic vocabulary of rights, entitlements, social provisions, community, social responsibility, living wage, job security, equality, and justice seem oddly out of place in a country in which the promise of democracy has been replaced by casino capitalism— a winner-take-all philosophy suited to lotto players and day traders alike. The ever-present corporate culture is reinforced by a pervasive fear and insecurity about the present, and a deep-seated skepticism in the public mind and worry that the future holds a more obscene version of the present. As the discourse of neoliberalism seizes the public imagination, there is no vocabulary for political or social transformation, democratically inspired visions, or critical notions of social agency to enlarge the meaning and purpose of democratic public life. Against the reality of low wage jobs, the erosion of social provisions for a growing number of people, and the expanding war against young people of color, the market-driven juggernaut continues to mobilize desires in the interest of producing market identities and market relationships that ultimately sever the link between education and social change while reducing agency to the obligations of consumerism.

Under such circumstances, citizens lose their public voice as market liberties replace civic freedoms and society increasingly depends on "consumers to do the work of citizens."[4] Moreover, as corporations become more deregulated and deterritorialized, the political state increasingly is transformed into the business state, and as Noreena Hertz observes, "Economics has become the new politics, and business is in the driving seat."[5] What is troubling is not simply that ideas associated with freedom and agency are defined through the prevailing ideology and principles of the market, which is the case; or that neoliberal ideology wraps itself in what appears to be unassailable common sense, which it attempts; or finally, that it prohibits or censors critics, which it simply can't. What is more worrisome is that in the face of all sorts of political chicanery, the populace seems to resist all nonmarket alternatives and is convinced of its own helplessness, and that there are no alternatives to the present. As Zygmunt Bauman notes, "What, however, makes the neo-liberal world-view sharply different from other ideologies—indeed, a phenomenon of a separate class—is precisely the absence of questioning; its surrender to what is seen as the implacable and irreversible logic of social reality."[6] Our critique is not simply aimed at the willingness of neoliberalism's exponents to make their own assumptions problematic. On the contrary, the very viability of politics itself is at stake, as formal and informal public spaces for educational exchange and debate atrophy or disappear altogether.

Within neoliberalism's market-driven discourse, corporate culture becomes both the model for the good life and the paradigmatic sphere for defining individual success and fulfillment. We use the term "corporate culture" to refer to an ensemble of ideological and institutional forces that functions politically and pedagogically to both govern organizational life through senior managerial control and to fashion flexible and compliant workers, depoliticized consumers, and passive citizens.[7] Citizenship is portrayed as an utterly solitary affair whose aim is to produce competitive, self-interested individuals vying for their own material and ideological gain. Corporate culture either cancels out or devalues social, class-specific, and racial injustices in the existing social order. It does so by absorbing the democratic impulses and practices of civil society within an appeal to market-based freedoms and narrow economic relations. Corporate culture becomes an all-encompassing source of market identities, values, and practices. The good life, in this discourse, "is construed in terms of our identities as consumers— we are what we buy."[8] For example, some neoliberal advocates argue that the health care and education crises faced by many states can be solved by selling off public assets to private interests. The Pentagon even considered, if only for a short time, turning the War on Terror and security concerns over to futures markets. Thus, public spheres are replaced by commercial spheres, as the substance of critical democracy is emptied out and replaced by a democracy of markets, goods, services, and the increasing expansion of the cultural and political power of corporations throughout the world.

Accountable only to the bottom line of profitability, corporate culture has signaled a radical shift in the notion of public culture, the meaning of citizenship, and the defense of the public interest. The rapid resurgence of corporate power in the last 20 years and the attendant reorientation of culture to the demands of commerce and deregulation have substituted the language of personal responsibility and private initiative for the discourses of social responsibility and public service. This can be seen in the enactment of government policies designed to dismantle state protections for the poor, the environment, working people, and people of color.[9] For example, the 2003 federal budget enacted by President George W. Bush and the Republican-dominated Congress eliminated 8,000 homeless kids from educational benefits, terminated child care to 33,000 children, and cut 500,000 young people from after-school programs.[10] At the same time, half a million poor families and their children will be dropped from receiving any heating assistance. Moreover, this budget allocates more money for tax cuts for the rich than it does for education and low-income child care combined.[11] That the Bush administration places a low priority on investing in education, in spite of claims to the contrary, can be seen in the fact that the slated federal budget for 2004 allocates $308.5 billion to the Pentagon and only $34.7 billion to education.[12]

Unchecked by traditional forms of state power and removed from any sense of place-based allegiance, global neoliberal capitalism appears more detached than ever from traditional forms of political power bounded by nations and ethical considerations of people in specific localities. Public-sector activities such as transportation (in spite of the Amtrak bailout, which is as an exception to the rule), health care, and education are no longer safeguarded from incursions by the buying-and-selling logic of the market. As we write this, the Republican-controlled House of Representatives has passed a bill that will subsidize private plans competing with medicare, a tactic "clearly intended to undermine medicare over time."[13] The consequences are evident everywhere, but especially visible in the university where the language of the corporate commercial paradigm describes students as customers, college admissions as "closing a deal," and university presidents as CEOs.[14] But there is more at stake here than simple linguistic shifts that signal the commodification of everyday life. There is, as Pierre Bourdieu has argued, the emergence of a Darwinian world marked by the ongoing atrophy of autonomous spheres of cultural production such as journalism, academic publishing, and film; the destruction of collective structures capable of counteracting the widespread imposition of commercial values and effects of the pure market; the creation of a global reserve army of the unemployed; and the subordination of nation-states to the real masters of the economy.[15]

We are not suggesting that market institutions and investments cannot at times serve public interests, but rather that in the absence of vibrant, democratic public spheres, unchecked corporate power respects few boundaries based on self-restraint and the greater public good, and is increasingly unresponsive to those broader human values that are central to a democratic civic culture. We believe that at this point in American history, neoliberal capitalism is not simply too overpowering, but also that "democracy is too weak."[16] Hence, we witness the increasing influence of money over politics, corporate interests overriding public concerns, and the growing tyranny of unrestrained corporate power and avarice refashioning education at all levels. The economist Paul Krugman recently described a cultural revolution of values afoot in American life equal to that of the sexual revolution—one that reflects a neo-Darwinian ethic that shows no concern for the widening of already vast inequalities between rich and poor, black and white. Increasing evidence of the shameless greed-is-good ethos is visible in the corruption and scandals that have rocked giant corporations such as Enron, WorldCom, Xerox, Tyco, Walmart, and Adelphia. The fall-out suggests a widening crisis of leadership as United States economic interests increasingly dictate world trade policy. Guido Rossi, a former Italian Telecom chairman, points out that "What is lacking in the U.S. is a culture of shame. No C.E.O. in the U.S. is considered a thief if he does something wrong. It is a kind of moral cancer."[17] And indeed, one would be hard-pressed to find the kind of outrage directed toward corporate criminals to equal the venom spat at stereotypical images of young black women on welfare— so powerful is the identification with the wealthy, so complete is the demonization of the poor. Clearly, there is more at stake in this crisis than simply the rapacious greed of a few high-profile CEOs; there is the historic task of challenging neoliberalism and market fundamentalism as we attempt to reassert the meaning of democracy, citizenship, social justice, and civic education.

Struggling for substantive democracy is both a political and educational task. Fundamental to the health of a vibrant democratic culture is the recognition that education must be treated as a public good—a crucial site where students gain a public voice and come to grips with their own power and responsibility as social agents. Higher education (as well as public education) cannot be viewed merely as a commercial investment or a private good based exclusively on career-oriented needs. Reducing higher education to the handmaiden of corporate culture works against the critical social imperative of educating citizens who can sustain and develop inclusive democratic public spheres. Lost in the merging of corporate culture and higher education is a historic and honorable democratic tradition that extends from John Adams to W. E. B. Du Bois to John Dewey, one that we have mentioned repeatedly throughout this book, that extols the importance of education as essential for a democratic public life.[18] Education within this tradition integrated knowledge and civic values necessary for independent thought and individual autonomy with the principles of social responsibility. Moreover, it cast a critical eye on the worst temptations of profit-making and market-driven values. For example, Sheila Slaughter has argued persuasively that at the close of the nineteenth century, "professors made it clear that they did not want to be part of a cutthroat capitalism. . . . Instead, they tried to create a space between capital and labor where [they] could support a common intellectual project directed toward the public good."[19] Amherst College President Alexander Meiklejohn echoed this sentiment in 1916 when he suggested:

> Insofar as a society is dominated by the attitudes of competitive business enterprise, freedom in its proper American meaning cannot be known, and hence, cannot be taught. That is the basic reason why the schools and colleges, which are presumably commissioned to study and promote the ways of freedom, are so weak, so confused, so ineffectual.[20]

As the line between for-profit and not-for-profit institutions of higher education collapses, educator John Palattela observes, many "schools now serve as personnel offices for corporations"[21] and quickly dispense with the historically burdened though important promise of creating democratic mandates for higher education. Not surprisingly, students are now referred to as "customers," while faculty are defined less through their scholarship than through their ability to secure funds and grants from foundations, corporations, and other external sources. Instead of concentrating on critical teaching "that prepares citizens for active participation in a democratic society"[22] and research aimed at promoting the public good, faculty are now urged to focus on corporate largesse. Rather than being esteemed as engaged teachers and rigorous researchers, faculty are now valued as multinational operatives and, like their corporate counterparts, increasingly vulnerable to the threat of fixed-term contracts and "flexibilization."

Competition for top faculty among colleges and universities are now described in terms once appropriate for Hollywood celebrities. The *Boston Globe* recently ran a story in which faculty were ranked according to star power. Some faculty not only bought into this grotesque description of their vocation, but proved quite blunt about what motivates their job choices. For instance, one sought-after alleged "star"and cheerleader for a robust American empire, Niall Ferguson, stated unabashedly that what finally convinced him to take a job at New York University was the allure of money and power. Repeating an exchange with the university president, he fills in the details without the slightest hint of shame or embarrassment: "'Niall, you're interested in money and power, right?' I said, 'Yes.' And he said, 'Well, why don't you come and work where the money and power are?'"[23] Tragically, the cost of such celebrity faculty who rarely teach undergraduates gets passed on to these students nonetheless, in the form of spiraling tuition rates.

Such rhetoric reflects a fundamental shift in how we think about the relationship among educators, corporate culture, and democracy.[24] One of the most important indications of such a change can be seen in the ways in which educators are being asked to rethink the role of higher education and their place within it. We believe that the struggle to reclaim higher education must be seen as part of a broader battle over the defense of public goods. At the heart of such a struggle is the need to challenge the ever-growing discourse and influence of neoliberal corporate power and corporate politics. We also want to offer some suggestions as to what educators can do to reassert the primacy of higher education as an essential sphere for expanding and deepening the processes of democracy and civil society.

The Business of Higher Education

When the market interests totally dominate colleges and universities, their role as public agencies significantly diminishes—as does their capacity to provide venues for the testing of new ideas and the agendas for public action. What is lost is the understanding that knowledge has other than instrumental purposes, that ideas are important whether or not they confer personal advantage.

—Robert Zemsky[25]

Higher education, for many educators, as we mentioned in chapter 4, is a place of public purpose, a central site for keeping alive the tension between market values and those values of civil society that cannot be measured in narrow commercial terms but are crucial to an inclusive and nonrepressive democracy. Education must not be confused with training—suggesting all the more the role that educators might play in preventing the private sector from hijacking the purpose and mission of higher education in the interests of producing a flexible and docile workforce. Educators as different as Robert Zemsky and Derek Bok have raised disturbing questions about the growing commercialization of higher education and its willingness to define itself largely as a consumer good.[26] Critical citizens aren't born, they're made, and unless citizens are critically educated and well-informed, democracy is doomed to failure. Unfortunately, as Richard Ohmann observes, the damaged, though important, civic mission of higher education is increasingly being replaced by the goals and values of the corporate university, which attempts to define all knowledge, values, and activity in terms of the marketplace. The corporate university, according to Ohmann,

> acts like a profit-making business rather than a public or philanthropic trust. Thus, we hear of universities applying productivity and performance measures to teaching (Illinois); of plans to put departments in competition with one another for resources (Florida); of cutting faculty costs not only by replacing full-timers with part-timers and temps and by subcontracting for everything from food services to the total management of physical plant, but also by substituting various schemes of computerized instruction; and so on.[27]

The growing influence of corporate culture on university life in the United States has served largely to undermine the distinction between higher education and big business that many educators want to preserve. Laboring under massive budget cuts, universities are turning to corporations to provide needed funding. The consequences, however, are troubling. Collaborative relationships among faculty suffer as some firms insist that the results of corporate-sponsored research be kept secret. In other cases, researchers funded by corporations have been prohibited from speaking about their research at conferences, talking on the phone with colleagues, or making their labs available to faculty and students not directly involved in the research. Derek Bok reports that "Nearly one in five life-science professors admitted that they had delayed publication by more than six months for commercial reasons."[28] Equally disturbing is the growing number of academics who either hold stocks or other financial incentives in the companies sponsoring their research and the refusal on the part of many universities to institute disclosure policies that would reveal such conflicts of interest.[29] Moreover, as the boundaries between public and commercial values become blurred, many academics appear less as disinterested truth seekers than as apologists for corporate values and profiteering. This becomes particularly startling with respect to corporate-funded medical research.

The New England Journal of Medicine reported recently that "medical schools that conduct research sponsored by drug companies routinely disregard guidelines intended to ensure that the studies are unbiased and that the results are shared with the public."[30] The medical schools did very little to minimize the effect of corporate influence in medical research. The Journal of the American Medical Association also has reported recently that "one fourth of biomedical scientists have financial affiliations with industry . . . and that research financed by industry is more likely to draw commercially favorable conclusions."[31] Corporate power and influence also shapes the outcome of the research and the design of the clinical trials. Hence, it is not surprising to find, as the journal reported, that "studies reported by the tobacco industry reported pro-industry results [and that] studies on pharmaceuticals were affected by their source of funds as well."[32] In some instances, corporations place pressure on universities to suppress the publication of those studies whose data questions the effectiveness of their wares, threatening not only academic integrity but also public health and safety. For example, Canada's largest pharmaceutical company, Apotex, attempted to suppress the findings of a

University of Toronto researcher, Dr. Nancy Olivieri, when she argued that the "drug the company was manufacturing was ineffective, and could even be toxic."[33] The University of Toronto not only refused to provide support for Dr. Olivieri, it also suspended her from her administrative role as program director, and warned her and her staff not to talk publicly about the case. It was later disclosed that "the university and Apotex had for some years been in discussions about a multimillion-dollar gift to the university and its teaching hospitals."[34]

As corporate culture and values shape university life, corporate planning replaces social planning, management becomes a substitute for leadership, and the private domain of individual achievement replaces the discourse of participatory politics and social responsibility. While it is difficult to predict what the eventual consequences might be, Derek Bok argues that university leaders have not paid enough attention to this trend. He predicts that if the commercialization of higher education is not brought under control, the institution could end up cheapened and trivialized. He writes:

> One can imagine a university of the future tenuring professors because they bring in large amounts of patent royalties and industrial funding; paying high salaries to recruit "celebrity" scholars who can attract favorable media coverage; admitting less than fully qualified students in return for handsome parent gifts; soliciting corporate advertising to underwrite popular executive programs; promoting Internet courses of inferior quality while canceling worthy conventional offerings because they cannot cover their costs; encouraging professors to spend more time delivering routine services to attract corporate clients, while providing a variety of symposia and "academic" conferences planned by marketing experts in their development offices to lure potential donors to campus.[35]

As the power of higher education is reduced in its ability to make corporate power accountable, it becomes more difficult for faculty, students, and administrators to address pressing social and ethical issues.[36] This suggests a perilous turn in American society, one that threatens both our understanding of democracy as fundamental to our basic rights and freedoms and the ways in which we can rethink and reappropriate the meaning, purpose, and future of higher education.

The Rise of the Academic Manager in Higher Education

In the aftermath of the terrorist attacks of September 11, 2001, and in the midst of the current recession, many colleges and universities are experiencing financial hard times. These circumstances have been exacerbated by an economic downturn brought about by the fiscal crisis of the states, exorbitant tax breaks for the wealthy matched by growing budget deficits, record-breaking unemployment rates, a soaring federal debt, and the enormous cost of maintaining the military occupation of Iraq (estimated at $5 billion a month), all of which have resulted in a sharp reduction of state aid to higher education. Rather than provide increased aid for colleges and universities (or unable to do so because of declining tax revenues), state legislators encourage tuition increases. Such approaches to rising costs in higher education not only punish students in the form of crippling debt or denied access—they simply do not work. As a result, many colleges and universities are all too happy to allow corporate leaders to run their institutions, form business partnerships, establish cozy relationships with business-oriented legislators, and develop curricular programs tailored to the needs of corporate interests.[37] Bill Gates, Jack Welch, Michael Milken, Warren Buffet, and other members of the Fortune 500 "club" continue to be viewed as educational prophets—in spite of the besmirched reputation of former CEOs such as Kenneth Lay of Enron, Al Dunlap of Sunbeam, and Dennis Kozlowski of Tyco.[38] And yet, the only qualifications they seem to offer is that they have been successful in accumulating huge amounts of money for themselves and their shareholders by laying off thousands of workers in order to cut costs and raise profits. For example, between 1990 and 2000, the average CEO salary rose 571 percent, while during the same period, the salary of the average worker rose 37 percent.[39] What exactly is the pedagogical role such high-profile profiteers are to bring to the "beloved community" of university scholars—what lessons on public service are they in a position to confer on students? While Gates, Milken, and others couch their alleged commitment to education in the rhetoric of public service, corporate organizations such as the Committee for Economic Development, an organization of about 250 corporations, have been more blunt about their educational concerns.[40] Not only has the group argued that social goals and services get in the way of learning basic skills, but that many employers in the business community feel dissatisfied because "a large majority of their new hires lack adequate writing and problem-solving skills."[41] Such skills are championed not because they form the basis for literacy itself, but because without them workers do not perform well.

Even when corporate CEOs take on the role of heading for-profit universities, they are quite open about both who they serve and how they feel about public values. For example, Ronald Tayler, the chief operating officer of DeVry University, the second-largest for-profit university in the United States, says, "The colossally simple notion that drives DeVry's business is that if you ask employers what they want and then provide what they want, the people you supply to them will be hired."[42] On the issue of the university's relationship to noncommercial values and the public good, John Sperling, the founder of the University of Phoenix, the largest for-profit university, says boldly, "I'm not involved in social reform."[43]

Corporate culture, in large measure, lacks a vision beyond its own pragmatic interests in profit and growth, seldom providing a self-critical inventory about its own ideology and its effects on public health, the environment, or the stability and gainful employment of citizens. It is difficult to imagine such concerns arising within corporations where questions of consequence begin and end with the bottom line. Clearly, neoliberal advocates, in the drive to create wealth for a limited few, have no incentives for taking care of basic social needs, or maintaining even the most minimal requirements of the social contract designed to provide a modicum of security and Safety for all Americans. This is obvious not only in their attempts to render the welfare state obsolete, privatize all public goods, and destroy traditional state-provided safety nets, but also in their disregard for the environment, misallocation of resources between the private and public sectors, and relentless pursuit of profits. It is precisely this lack of emphasis on being a public servant and an academic citizen that is missing from the leadership models that corporate executives transfer to their roles as academic administrators. Unfortunately, it often pays off in financial benefits for the corporations, which are not accountable to the public interest.

As market-fund mogul George Soros has pointed out, neoliberal economic agendas promote a kind of market fundamentalism based on the untrammeled pursuit of self-interest—often wrapped up in the post–September 11 language of patriotism. In a post-9/11 world, some advertisers now surround their sales pitches with images of the flag, selling along with their commodities the supposition that consumerism is the essence of patriotism. Most advertising campaigns, however, make no appeal to redeeming human values, no matter how disingenuous. The distinguishing features of market fundamentalism are that "morality does not enter into [its] calculations" and it does not necessarily serve the common interest, nor is it capable of taking care of collective needs and ensuring social justice.[44] One egregious example of this type of advertisement can be seen in a television ad sponsored by Hotwire.com, a leading discount travel site. The ad begins with a father and son on a diving board. The father is trying to teach the boy how to dive. The son suddenly turns to his dad and points to a quarter in the pool. The father eyes the quarter, ruthlessly pushes the kid off the diving board and plunges into the water to retrieve it. In the next shot, the dad is standing in the pool, triumphant, one hand above his head holding the quarter up for the viewer to marvel at the retrieved prize. The ad ends extolling the dad as its "kind of customer" (cheap)—apparently finding it appropriate to use an act of child abuse for satirical fodder, and indifferent to the kind of selfish character they are extolling.

What society allows this kind of child abuse to be served up for a good laugh, or for that matter to be even presented on the national media? In this climate, it is highly unlikely that corporations such as Disney, IBM, or General Motors will seriously address the political and social consequences of the policies they implement, which have resulted in downsizing, deindustrialization, and the "trend toward more low-paid, temporary, benefit-free, blue- and white-collar jobs and fewer decent permanent factory and office jobs."[45] Clearly, the interests served by such changes, as well as the consequences they have for working people, immigrants, and others, detract from those democratic arenas that business seeks to "restructure." Mega-corporations will say nothing about their profound role in promoting the flight of capital abroad; the widening gap between intellectual, technical, and manual labor; the growing class of those permanently underemployed in a mass of "deskilled jobs"; the increasing inequality between the rich and the poor; or the scandalous use of child labor in Third World countries. Nor will they say anything critical about the control of the media by a handful of corporations and the effects of this concentration of power in undermining an effective system of political communication, which is crucial to creating an informed and engaged citizenry.[46] Rather, the onus of responsibility is placed on educated citizens to recognize that corporate principles of efficiency, accountability, and profit maximization have not created new jobs but in most cases have eliminated them.[47] It is citizens' responsibility to recognize that the world presented to them through allegedly objective reporting is mediated—and manipulated—by a handful of global media industries run by moguls such as Rupert Murdoch and Michael Eisner, though most Americans have little access to informed public debate or alternative viewpoints. Our point, of course, is that such omissions in public discourse constitute a defining principle of corporate ideology, which refuses to address—but must be made to address—the absence of moral vision in such calls for educational changes modeled after corporate management and ideology.

In the corporate model, knowledge is privileged as a form of investment in the economy, but appears to have little value in terms of self-definition, social responsibility, or the capacities of individuals to expand the scope of freedom, justice, and democracy.[48] Stripped of ethical and political considerations, knowledge offers limited (if any) insights into how schools should educate students to push against the oppressive boundaries of gender, class, race, and age domination. Nor does such a corporate language provide the pedagogical conditions for students to think critically, take risks politically, or imagine a world governed by civic values rather than corporate interests. Education is a moral and political practice and always embodies particular views of social life, a particular rendering of what community is, and an idea of what the future might hold. As such, the problems with American schools cannot be reduced to matters of accountability or cost-effectiveness. Nor can the solution to such problems be reduced to the spheres of management, economics, and technological quick fixes such as "distance education," which offers academic courses on-line. The problems of higher education must be addressed in terms of values and politics, while engaging critically the most fundamental beliefs Americans have as a nation regarding the meaning and purpose of education and its relationship to democracy.

Faculty and Students in the Corporate University

As universities increasingly model themselves after corporations, it becomes crucial to understand how the principles of corporate culture have altered the meaning and purpose of the university, the role of knowledge production in the twenty-first century, and the social practices inscribed within teacher–student relationships. The signs are not encouraging. Knowledge with a high market value is what counts, while those fields, such as the fine arts and humanities, that cannot be quantified in such terms will either be downsized or allowed to become largely irrelevant in the hierarchy of academic knowledge. Moreover, those professors who are rewarded for bringing in outside money will be more heavily represented in fields such as science and engineering, which attract corporate and government research funding. As Sheila Slaughter observes, "Professors in fields other than science and engineering who attract funds usually do so from foundations which account for a relatively small proportion of overall research funding."[49]

In other quarters of higher education, the influence of corporate culture can be seen not only in the refusal of political leaders to address the public purposes of colleges and universities, but also in attempts on the part of many politicians to align higher education with market-based ideologies. One telling example took place recently when Governor Mitt Romney put forth a plan to reorganize higher education in Massachusetts. The initiative, which has since been voted down by the legislature, would have reorganized the Amherst campus into a prestigious, independent research-based institution, privatized three public colleges, and merged the remainder into regional groupings so as to better serve the needs of distinct business niches. The overall result would be to split the Massachusetts system of higher education, establish a three-tier system of education designed to provide a quality education to an upper-middle-class elite, and to offer educational training to those economically disadvantaged students the system has served traditionally. William M. Bulger, the former president of the University of Massachusetts system, "blasted the proposal as an elitist 'corporate takeover' of higher education" and defined Romney's view of education as "nothing more than job training."[50]

There is more at stake in university reform than the principles of profit-making, the career needs of students, and the harsh realities of cost-cutting. Neoliberalism, fueled by its unwavering belief in market values and the unyielding logic of corporate profit-making, has little patience with noncommodified knowledge or with the more lofty ideals that have defined higher education as a public service. Romney's animosity toward educators and students alike is simply a more extreme example of the forces at work in the corporate world that would like to take advantage of the profits to be made in higher education, while simultaneously refashioning colleges and universities in the image of the new multinational conglomerate landscape. The corporate model fails to recognize that the public mission of higher education implies that knowledge has a critical function; that intellectual inquiry that is unpopular or debunking should be safeguarded and treated as an important social asset; and that faculty in higher education are more than merely functionaries of the corporate order. Such ideals are at odds with the vocational function that corporate advocates such as Romney want to assign to higher education.

While corporate values such as efficiency and downsizing in higher education appear to have caught the public's imagination at the moment, in fact such "reorganization" has been going on for some time. More professors are working part-time and at two-year community colleges than at any other time in the country's recent history. A 2001 report by the National Study of Postsecondary Faculty pointed out that "in 1998–1999, less than one-third of all faculty members were tenured. . . . [and that] in 1992–1993, 40 percent of the faculty was classified as part-time and in

1998–1999, the share had risen to 45 percent."[51] The American Council of Education reported in 2002 that "The number of part-time faculty members increased by 79 percent from 1981 to 1999, to more than 400,000 out of a total of one million instructors over all," and that the "biggest growth spurt occurred between 1987 and 1993, when 82 percent of the 120,000 new faculty members hired during that period were for part-time positions."[52] Creating a permanent underclass of part-time professional workers in higher education is not only demoralizing and exploitative for those who have such jobs; it also increasingly de-skills both part- and full-time faculty by increasing the amount of work they have to do. With less time to prepare, larger class loads, almost no time for research, and excessive grading demands, many adjuncts run the risk of becoming demoralized, ineffective, and unable to keep apace with new knowledge in their disciplines—let alone produce innovative research.

As power shifts away from the faculty to the administrative sectors of the university, adjunct faculty increase in number while effectively being removed from the faculty governance process. In short, the hiring of part-time faculty to minimize costs simultaneously maximizes managerial control over faculty and the educational process itself. As their ranks are depleted, full-time faculty live under the constant threat of being either given heavier workloads or of having their tenure contracts eliminated or drastically redefined through "post-tenure reviews." These structural and ideological factors send a chill through post-secondary faculty and undermine the collective power academics need to challenge the increasingly corporate-based, top-down administrative structures that are becoming commonplace in many colleges and universities.

The turn to downsizing and de-skilling faculty is also exacerbated by the attempts on the part of many universities to expand into the profitable market of distance education, whose on-line courses reach thousands of students. Such a market is all the more lucrative since it is being underwritten by the combined armed services, which in August of 2000 pledged almost $1 billion to "provide taxpayer-subsidized university-based distance education for active-duty personnel and their families."[53] David Noble has written extensively on the restructuring of higher education under the imperatives of the new digital technologies and the move into distance education. If he is correct, the news is not good. According to Noble, on-line learning largely functions through pedagogical models and methods of delivery that not only rely on standardized, prepackaged curricula and methodological efficiency; they also reinforce the commercial penchant toward training, de-skilling, and de-professionalization. The de-skilling of the professoriate will further fuel the rise in the use of part-time faculty, who will be "perfectly suited to the investor-imagined university of the future."[54]

Columbia University's Teachers College president, Arthur Levine, has predicted that the new information technology may soon make the traditional college and university obsolete. He is hardly alone in believing that on-line education will either radically alter or replace traditional education. As journalists Eyal Press and Jennifer Washburn point out, "In recent years academic institutions and a growing number of Internet companies have been racing to tap into the booming market in virtual learning."[55] The marriage of corporate culture, higher education, and the new high-speed technologies also offers universities big opportunities to cut back on maintenance expenses, eliminate entire buildings such as libraries and classrooms, and trim labor costs.

Universities and colleges across the country are flocking to the on-line bandwagon. As Press and Washburn point out, "more than half of the nation's colleges and universities deliver some courses over the Internet."[56] Mass-marketed degrees and courses are not only being offered by prestigious universities such as Seton Hall, Stanford, Harvard, the New School, and the University of Chicago; they are also giving rise to cyber-backed colleges such as the Western Governors University and for-profit, stand-alone, publicly traded institutions such as the University of Phoenix. We are not suggesting that technologies cannot improve classroom instruction, ameliorate existing modes of communication, or simply make academic work more interesting. The real issue is whether such technology in its various pedagogical uses in higher education is governed by a technocratic rationality that undermines human freedom and democratic values. As Herbert Marcuse has argued, when the rationality that drives technology is instrumentalized and "transformed into standardized efficiency . . . liberty is confined to the selection of the most adequate means for reaching a goal which [the individual] did not set."[57] The consequence of the substitution of technology for pedagogy is that instrumental goals replace ethical and political considerations, to the detriment of classroom control by teachers and in favor of standardization and rationalization of course materials. Zygmunt Bauman underscores such a danger by arguing that when technology is coupled with calls for efficiency modeled on instrumental rationality, it almost always leads to forms of social engineering that seem increasingly "reasonable" and dehumanizing at the same time.[58] In other words, when the new computer technologies are tied to narrow forms of instrumental rationality, they serve as "moral sleeping pills," which are increasingly made available by corporate power and the modern bureaucracy of higher education.[59] The issue here is not only that the new computer technologies enable on-line pedagogical approaches such as distance education and supplant place-based, "real" education with limited forms of simulated and virtual exchanges, but that such technologies, when not shaped by ethical considerations, collective debate, and dialogical approaches, lose whatever potential they might have for linking education to critical thinking and learning to democratic social change.[60] Under such conditions, the new technologies run the risk of contributing to the de-skilling of teachers, the growth in a reserve army of part-time instructors, and a dehumanizing pedagogy for students.

In fact, when business concerns about efficiency and cost-effectiveness replace the imperatives of critical learning, a division based on social class begins to appear. Poor and marginalized students will get low-cost, low-skilled knowledge and second-rate degrees from on-line sources, while those students being educated for leadership positions in the elite schools will get personalized instruction and socially interactive pedagogies in which high-powered knowledge, critical thinking, and problem-solving will be a priority (coupled with a high-status degree). Under such circumstances, traditional modes of class and racial tracking will be reinforced and updated in what David Noble calls "digital diploma mills."[61] Noble underemphasizes, in his otherwise excellent analysis, indications that the drive toward corporatizing the university will take its biggest toll on those second- and third-tier institutions that are increasingly defined as serving no other function than to train semi-skilled and obedient workers for the new postindustrial order. The role slotted for these institutions is driven less by the imperatives of the new digital technologies than by the need to reproduce a gender, racial, and class division of labor that supports the neoliberal global market revolution and its relentless search for bigger profits.

Held up to the profit standard, universities and colleges will increasingly calibrate supply to demand, and the results look ominous with regard to what forms of knowledge, pedagogy, and research will be rewarded and legitimated. As colleges and corporations collaborate over the content of degree programs, particularly with regard to on-line graduate degree programs, college curricula run the risk of being narrowly tailored to the needs of specific businesses. For example, Babson College developed a master's degree program in business administration specifically for Intel workers. Similarly, the University of Texas at Austin is developing an on-line master of science degree in science, technology, and commercialization that caters only to students who work at IBM. Moreover, the program will orient its knowledge, skills, and research to focus exclusively on IBM projects.[62] Not only do such courses come dangerously close to becoming company training workshops; they also open up higher education to powerful corporate interests that have little regard for the more time-honored educational mandate to cultivate an informed, critical citizenry.

While it is crucial to recognize the dangers inherent in on-line learning and the instructional use of information technology, it is also important to recognize that there are many thoughtful and intelligent people who harness such technologies in ways that can be useful for educators and students. We do not want to suggest that on-line distance education is the most important or only way in which computer-based technologies can be used in higher education, or that the new electronic technologies by default produce oppressive pedagogical conditions. Moreover, not everyone who uses these technologies can be simply dismissed as living in a middle-class world of techno-euphoria in which computers are viewed as a panacea. Andrew Feenberg, a professor at San Diego State University and a former disciple of Herbert Marcuse, rejects the essentialist view that technology reduces everything to functions, efficiency, and raw materials, "while threatening both spiritual and material survival."[63] Feenberg argues that the use of technology in both higher education and other spheres has to be taken up as part of a larger project to expand democracy, and that under such conditions it can be used "to open up new possibilities for intervention."[64] Many educators use e-mail, the Internet, on-line discussion groups, and computer-based interaction to provide invaluable opportunities for students to gain access to new knowledge and to enhance communication, dialogue, and learning. But with this caveat in mind, there is still the important question of how technology might threaten the integrity of democratic education, identities, values, and institutions. This question returns us to some more critical considerations.

On-line courses also raise important issues about intellectual property—who owns the rights for course materials developed for on-line use. Because of the market potential of on-line lectures and course materials, various universities have attempted to lay ownership claims to such knowledge. The passing of the 1980 Bayh-Dole Act and the 1984 Public Law 98-620 by the United States Congress enabled "universities and professors to own patents on discoveries or inventions made as a result of federally supported research."[65] These laws accorded universities intellectual property rights, with specific rights to own, license, and sell their patents to firms for commercial profits. The results have been far from unproblematic.[66] Julia Porter Liebeskind, a professor at the Marshall School of Business, points to three specific areas of concern that are worth mentioning.

First, the growth of patenting by universities has provided a strong incentive "for researchers to pursue commercial projects," especially in light of the large profits that can be made by faculty.[67] For instance, five faculty members at the University of California system and an equal number at Stanford University in 1995 earned a total of $69 million in licensing income (fees and royalties). And while it is true that the probability for large profits for faculty is small, the possibility for high-powered financial rewards cannot be discounted in the shaping of the production of knowledge and research at the university.

Second, patenting agreements can place undue restraints on faculty, especially with respect to keeping their research secret and delaying publications, or even prohibiting "publication of research altogether if it is found to have commercial value."[68] Such secrecy undermines faculty collegiality and limits a faculty member's willingness to work with others; it also damages faculty careers and prevents significant research from becoming part of the public intellectual commons. Derek Bok concisely sums up some of the unfortunate consequences, particularly in the sciences, that plague higher education's complicity with the corporate demand for secrecy:

> It disrupts collegial relationships when professors cannot talk freely to other members of their department. It erodes trust, as members of scientific conferences wonder whether other participants are withholding information for commercial reasons. It promotes waste as scientists needlessly duplicate work that other investigators have already performed in secret for business reasons. Worst of all, secrecy may retard the course of science itself, since progress depends upon every researcher being able to build upon the findings of other investigators.[69]

Finally, the ongoing commercialization of research puts undue pressure on faculty to pursue research that can raise revenue and poses a threat to faculty intellectual property rights. For example, at the University of California at Los Angeles, an agreement was signed in 1994 that allowed an outside vender, On-lineLearning.net, to create and copyright on-line versions of UCLA courses. The agreement was eventually "amended in 1999 to allow professors' rights to the basic content of their courses . . . [but] under the amended contract, On-lineLearning retain[ed] their right to market and distribute those courses on-line, which is the crux of the copyright dispute."[70]

The debate over intellectual property rights calls into question not only the increasing influence of neoliberal and corporate values on the university, but also the vital issue of academic freedom. As universities make more and more claims on owning the content of faculty notes, lectures, books, computer files, and media for classroom use, the first casualty is, as Ed Condren, a UCLA professor points out, "the legal protection that enables faculty to freely express their views without fear of censorship or appropriation of their ideas."[71] At the same time, by selling course property rights for a fee, universities infringe on the ownership rights of faculty members by removing them from any control over how their courses might be used in the public domain.

As globalization and corporate mergers increase, new technologies develop, and cost-effective practices expand, there will be fewer jobs for certain professionals—resulting in the inevitable elevation of admission standards, restriction of student loans, and the reduction of student access to higher education, particularly for those groups who are marginalized because of their class and race.[72] Fewer jobs in higher education means fewer students will be enrolled, but it also means that the processes of vocationalization—fueled by corporate values that mimic "flexibility," "competition," or "lean production" and rationalized through the application of accounting principles— threaten to gut many academic departments and programs that cannot translate their subject matter into commercial gains. Programs and courses that focus on areas such as critical theory, literature, feminism, ethics, environmentalism, postcolonialism, philosophy, and sociology involve an intellectual cosmopolitanism or a concern with social issues that will be either eliminated or cut back because their role in the market will be judged as ornamental, or in the post-9/11 era, "unpatriotic," as we discussed in chapter 1. Similarly, those working conditions that allow professors and graduate assistants to comment extensively on student work, provide small seminars, spend time with student advising, conduct independent studies, and do collaborative research with both faculty colleagues and students do not appear consistent with the imperatives of downsizing, efficiency, and cost accounting.[73]

Students will also be affected adversely by the growing collaboration between higher education and the corporate banking world. As all levels of government reduce their funding to higher education, not only will tuition increase, but loans will increasingly replace grants and scholarships. Lacking adequate financial aid, students, especially poor students, will have to finance the high costs of their education through private corporations such as Citibank, Chase Manhattan, Marine Midland, and other lenders. According to the

U.S. Public Interest Group, student loans accounted for 20 percent of federal education assistance in 1976 but now have become the largest source of aid. The average student now graduates with a debt of more than $16,000, and one in three seniors have debts of more than $20,000.[74] As Jeff Williams points out, such loans "effectively indenture students for ten to twenty years after graduation and intractably [reduce] their career choices, funneling them into the corporate workforce in order to pay their loans."[75]

Of course, for many young people caught in the margins of poverty, low-paying jobs, recession, and "jobless recovery," the potential costs of higher education, regardless of its status or availability, will dissuade them from even thinking about attending college. Unfortunately, as state and federal agencies and university systems direct more and more of their resources (such as state tax credits and scholarship programs) toward middle- and upper-income students and away from need-based aid, the growing gap in college enrollments between high-income students (95 percent enrollment rate) and low-income students (75 percent enrollment rate) with comparable academic abilities will widen even further.[76] In fact, a recent report by a federal advisory committee claimed that nearly 48 percent of qualified students from low-income families would not be attending college in the fall of 2002 because of rising tuition charges and a shortfall in federal and state grants. The report claimed that "Nearly 170,000 of the top high-school graduates from low- and moderate-income families are not enrolling in college this year because they cannot afford to do so."[77] It also predicted that if the financial barriers that low- and moderate-income students face are not addressed, more than 2 million students by the end of the decade will not attend any form of higher education.[78]

Those students who enter higher education will often find themselves in courses being taught by an increasing army of part-time and adjunct faculty. Given personnel costs—"of which salaries and benefits for tenured faculty . . . typically account for 90 percent of operating budgets"[79]—university administrators are hiring more part-time faculty and depleting the ranks of tenured faculty. Applying rules taken directly from the cost-effective, downsizing strategies of industry, universities continuously attempt to cut budgets, maximize their efficiency, and reduce the power of the professoriate by keeping salaries as low as possible, substituting part-time teaching positions for full-time posts, chipping away at or eliminating employee benefits, and threatening to restructure or eliminate tenure. Not only do such policies demoralize the full-time faculty, exploit part-time workers, and overwork teaching assistants—they also cheat students. Too many undergraduates find themselves in oversized classes taught by faculty who are over-burdened by heavy teaching loads. Understandably, such faculty have little loyalty to the departments or universities in which they teach, rarely have the time to work collaboratively with other faculty or students, have almost no control over what they teach, and barely have the time to do the writing and research necessary to keep up with their fields of study. The result often demeans teachers' roles as intellectuals, proletarianizes their labor, and short-changes the quality of education that students deserve.[80]

We are not suggesting, of course, that the part-time workers are as deficient as the conditions they are forced to work under. It is one thing to be the victim of a system built on greed and scandalous labor practices, and another thing to take the heat for trying to make a living in such contexts. The real issue here is that these conditions are exploitative and the solutions for fixing the problem lie not simply in hiring more full-time faculty, but, as Cary Nelson points out, in reforming "the entire complex of economic, social and political forces operating on higher education."[81]

Neoliberalism's obsession with spreading the gospel of the market and the values of corporate culture has utterly transformed the nature of educational leadership, the purpose of higher education, the work relations of faculty, the nature of what counts as legitimate knowledge, and the quality of pedagogy itself. It has also restructured those spaces and places in which students spend a great deal of time outside of classrooms. Increasingly, corporations are joining up with universities to privatize a seemingly endless array of services that universities once handled by themselves. University bookstores are now run by corporate conglomerates such as Barnes & Noble, while companies such as Sodexho-Marriott (also a large investor in the U.S. private prison industry) run a large percentage of college dining halls, and McDonald's and Starbucks occupy prominent locations on the student commons. Student identification cards are now adorned with MasterCard and Visa logos, providing them with an instant line of credit. In addition, housing, alumni relations, health care, and a vast range of other services are now being leased out to private firms to manage and run. One consequence is that spaces once marked as public and noncommodified—spaces for quiet study or student gatherings—now have the appearance of a shopping mall. As David Trend points out:

> student union buildings and cafeterias took on the appearance—or were conceptualized from the beginning—as shopping malls or food courts, as vendors competed to place university logos on caps, mugs, and credit cards. This is a larger pattern in what has been termed the "Disneyfication" of college life. . . . a pervasive impulse toward infotainment . . . where learning is "fun," the staff "perky," where consumer considerations dictate the curriculum, where presentation takes precedence over substance, and where students become "consumers."[82]

Commercial logos, billboards, and advertisements now plaster the walls of student centers, dining halls, cafeterias, and bookstores. Everywhere students turn outside of the university classroom, they are confronted with vendors and commercial sponsors who are hawking credit cards, athletic goods, soft drinks, and other commodities that one associates with the local shopping mall. Universities and colleges compound this marriage of commercial and educational values by signing exclusive contracts with Pepsi, Nike, Starbucks, and other contractors, further blurring the distinction between student and consumer. The message to students is clear: customer satisfaction is offered as a surrogate for learning, "to be a citizen is to be a consumer, and nothing more. Freedom means freedom to purchase."[83] But colleges and universities do not simply produce knowledge and values for students, they also play an influential role in shaping their identities. If colleges and universities are to define themselves as centers of teaching and learning vital to the democratic life of the nation, they must acknowledge the real danger of becoming mere adjuncts to big business, or corporate entities in themselves. At the very least, this demands that university administrators, academics, students, and others exercise the political, civic, and ethical courage needed to refuse the commercial rewards that would reduce them to simply another brand name or corporate logo.

Does Higher Education Have a Democratic Future?

> What I defend above all is the possibility and the necessity of the critical intellectual. . . . There is no genuine democracy without genuine opposing critical powers.[84]

We want to return to the argument that corporations guided by the dictates of rapacious neoliberalism have been given too much power in this society, and that educators need to address this threat to all facets of public life organized around noncommodified principles such as the pursuit of knowledge, justice, freedom, and equality. Against the current drive to corporatize higher education, higher education needs to be safeguarded as a public good against ongoing attempts to organize and run it like a business. Rather than being viewed as a source of profits, in which curriculum becomes a commodity, students are treated as consumers and trained as workers, and faculty are relegated to the status of contract employees,[85] higher education should, at the very least, be embraced as a democratic sphere because it is one of the few public spaces left where students can learn to think for themselves, question authority, recover the ideals of engaged citizenship, reaffirm the importance of the public good, and expand their capacity to make a difference in society. Central to such a task is the challenge to resist the university becoming what literary theorist Bill Readings has called a consumer-oriented corporation more concerned about accounting than accountability, and whose mission, defined largely through an appeal to excellence, is comprehended almost exclusively in terms of a purely instrumental efficiency.[86]

The crisis of higher education, then, needs to be analyzed in terms of wider economic, political, and social forces that exacerbate tensions between those who value such institutions as democratic public spheres and those advocates of neoliberalism who see market culture as a master design for all human affairs. Educators must challenge all attempts to evacuate democracy of its substantive ideals by reducing it to the imperatives of hyper-capitalism and the glorification of financial markets. This requires, as Jeff Williams points out, that educators "distinguish the university as a not-for profit institution, which serves a public interest, from for-profit organizations, which by definition serve private interests and often conflict with public interests"; he goes on to suggest that they propose "new images or fictions of the university, to reclaim the ground of the public interest, and to promote a higher education operating in that public interest."[87]

The task of revitalizing such a public dialogue suggests that faculty, students, and administrators will have to create enclaves of resistance to question official forms of authority, increasingly standardized curricula, admissions policies that favor white, upper-middle-class students, classroom pedagogies that restrict student participation, and hiring policies that exploit graduate students and adjunct faculty. Beyond opening up spaces for critical analysis, educators must work together to highlight and critically evaluate the relationship between civil society and corporate power while simultaneously struggling to prioritize citizen rights over consumer rights.

But more is needed than defending higher education as a vital sphere in which to develop the proper balance between democratic ideals and market-based values. Given the current assault by politicians, conservative foundations, and the right-wing media on educators who have spoken critically about U.S. foreign policy in light of the tragic events of September 11 and the invasion of Iraq, it is politically crucial that educators at all levels of involvement in the academy be defended as intellectuals who provide a significant service to the nation, particularly in their attempts to exercise and protect academic freedom. Such an appeal cannot be made in the name of professionalism, but in terms of the civic good such intellectuals provide. As we have said throughout this book, too many academics have retreated into narrow specialties that serve largely to consolidate authority rather than critique its abuses. Refusing to take positions on controversial issues or to examine the role they might play in lessening human suffering, professionalized academics become models of moral indifference and civic spectatorship, unfortunate examples of what it means to disconnect learning from public life. On the other hand, many left and liberal academics have done little better, retreating into arcane discourses that offer them mostly the safe ground of the professional recluse. Making almost no connections to audiences outside of the academy or to the issues that bear down on their everyday lives, these academics have become largely irrelevant. This is not to suggest that they do not publish or speak at symposiums, but that they often do so to very limited audiences and in a language that is often overly abstract, highly aestheticized, rarely takes an overt political position, and seems largely indifferent to broader public issues.

Engaged intellectuals such as Noam Chomsky, Susan Sontag, Howard Zinn, Barbara Ehrenreich, Robert McChesney, Ellen Willis, Stanley Aconowitz, and the late Edward Said and Pierre Bourdieu have offered a different and more committed role for academics. For instance, Noam Chomsky claims that "the social and intellectual role of the university should be subversive in a healthy society. . . . individuals and society at large benefit to the extent that these liberatory ideals extend throughout the educational system—in fact, far beyond."[88] Postcolonial and literary critic Edward Said takes a similar position and argues that academics should engage in ongoing forms of permanent critique of all abuses of power and authority, "to enter into sustained and vigorous exchange with the outside world," as part of a larger project of helping "to create the social conditions for the collective production of realist utopias."[89]

We believe that intellectuals who work in our nation's universities should represent the conscience of this society because they not only shape the conditions under which future generations learn about themselves and their relations to others and the world, but also because they engage in pedagogical practices that are by their very nature moral and political, rather than simply profit-maximizing and technical. At its best, such pedagogy bears witness to the ethical and political dilemmas that animate the broader social landscape; these approaches are important because they provide spaces that are both comforting and unsettling, spaces that both disturb and enlighten.

Pedagogy in this instance not only works to shift how students think critically about the issues affecting their lives and the world at large, but potentially energizes them to seize such moments as possibilities for acting on the world and engaging it as a matter of reclaiming politics and rethinking power in the interest of social justice. The appeal here is not merely ethical; it also addresses material resources, access, and policy decisions, while viewing power as crucial to any viable notion of individual and social agency.

Situated within the broader context of social responsibility, democratic politics, and the challenges of dignifying human life, higher education should be an institution that offers students the opportunity to involve themselves in the deepest problems of society and to acquire the knowledge, skills, and ethical vocabulary necessary for critical dialogue and broadened civic participation. This necessitates developing pedagogical conditions for students to come to terms with their own sense of power and public voice by enabling them to examine critically what they learn in the classroom "within a more political or social or intellectual understanding of what's going on" in their lives and the world at large.[90] In addition to addressing interdisciplinary modes of knowledge, students should be given the opportunity to take responsibility for their own ideas, take intellectual risks, develop a sense of respect for others, and learn how to think critically in order to function in a wider democratic culture. At issue here is providing students with an education that allows them to recognize the dream and promise of a substantive democracy, particularly the idea that as citizens they are "entitled to public services, decent housing, safety, security, support during hard times, and most importantly, some power over decision making."[91] But as we have stressed throughout this book, students also need to cross the boundary that separates colleges and universities from the larger world in ways that go beyond critical analysis or close textual readings. For instance, higher education should provide students with opportunities to use their knowledge and skills to engage in community service, organize partnerships between schools and nonprofits, or protest racism and poverty by actually challenging their manifestations within the larger community and social order. Civic duties might take the form of joining with groups to resist the growing criminalization of social problems such as drug use, poverty, and homelessness, and to find ways to help those minorities of class and color, especially young people, who desperately need job training, education, and literacy skills. They could assist those working poor who need health and child care, and help in dealing with welfare agencies, the courts, and other elements of the state. Academics and students alike could use their skills and institutional resources (in ways proportional to their access and abilities) to address the growing deterioration of the public schools, the crisis of unemployment among adults, and the literacy crisis in those rural and urban centers marked by poverty and a massive disparity of wealth. While it is important to define colleges and campuses as crucial public spheres, it is not enough if educators are to take seriously the link between learning, public values, and the principles of leadership. Learning should be viewed both as an individual process enabling maturity and autonomy and as a social practice capable of influencing and improving civic life. Learning and social criticism should be connected to forms of worldliness in which ideas are given meaning and agency is formed in the space of the public. Theory, like learning itself, cannot fully understand politics, social problems, issues, or public values without engaging them through the struggles in which they manifest themselves daily in the polity.

Organizing against the corporate takeover of higher education also means fighting to protect the jobs of full-time faculty, turning adjunct jobs into full-time positions, expanding benefits to part-time workers, and putting power into the hands of faculty and students. Moreover, such struggles must address the exploitative conditions under which many graduate students work, constituting a de facto army of service workers who are underpaid, overworked, and shorn of any real power or benefits.[92] Similarly, programs in many universities that offer remedial programs, affirmative action, and other crucial pedagogical resources are under massive assault, often by conservative trustees who want to eliminate from the university any attempt to address the deep inequities in society, while simultaneously denying a decent education to minorities of color and class. For example, City University of New York, as a result of a decision made by a board of trustees, decided to end "its commitment to provide remedial courses for academically unprepared students, many of whom are immigrants requiring language training before or concurrent with entering the ordinary academic discipline. . . . Consequently . . . a growing number of prospective college students are forced on an already overburdened job market."[93] Both teachers and students increasingly bear the burden of overcrowded classes, limited resources, and hostile legislators.

But, once again, resistance to neoliberal ideology and its onslaught against public goods, services, and civic freedoms cannot be limited either to the sphere of higher education or to outraged faculty. Educators and students should consider ways to join with community people and social movements around a common platform that resists the corporatizing of schools, the rollback in basic services even as tuition spirals out of control, and the exploitation of teaching assistants and adjunct faculty. There are several crucial lessons that faculty can learn from the growing number of broad-based student movements that are protesting neoliberal economic policies and the ongoing commercialization of the university and everyday life. Students from colleges across the United States and Canada have "held a series of 'teach-ins' challenging the increasing involvement of corporations in higher education."[94] Students from Yale, Harvard, Florida State University, and the University of Minnesota, among other schools, have organized debates, lectures, films, and speakers to examine the multifaceted ways in which corporations are affecting all aspects of higher education. Since the election of George W. Bush in 2000, the pace of such protests on and off campuses has picked up and spawned a number of student protest groups, including the United Students Against Sweatshops (USAS), with over 180 North American campus groups,[95] the nationwide 180/Movement for Democracy and Education, and a multitude of groups protesting the policies of the World Trade Organization and the International Monetary Fund, and massive demonstrations around the globe protesting the U.S. war with Iraq.[96]

Students have held hunger strikes, blocked traffic in protest of the brand-name society, conducted mass demonstrations against the WTO in Seattle and other cities, held peace rallies protesting the war in Afghanistan and Iraq, and demonstrated against the working conditions and use of child labor in the $2.5-billion collegiate apparel industry. They have also protested against tuition hikes and the Bush administration's attack on civil liberties through the passage of antiterrorist legislation. One of the common threads that ties them together is their resistance to the increasing incursion of corporate power on higher education and the growing militarization of public life.

Many students reject the model of the university as a business. They recognize that the corporate model of leadership fosters a narrow sense of responsibility, agency, and public values because it lacks a vocabulary attentive to matters of justice, equality, fairness, equity, and freedom—values crucial to a vibrant democratic culture. Students are refusing to be treated as consumers rather than as members of a university community in which they have a voice and have some say in how the university is organized and run. The alienation and powerlessness that ignited student resistance in the 1960s appears to be alive and growing today on college campuses across the country.

Student resistance to unbridled corporate power has also manifested itself outside of the campus in struggles for global justice that have taken place in cities such as Seattle, Prague, Washington, D.C., Davos, Porto Alegre, Melbourne, Quebec, Gothenburg, Genoa, and New York.[97] These anticorporate struggles not only include students, but also labor unions, community activists, environmental groups, and other social movements. These struggles offer students alliances with nonstudent groups, both within and outside the United States, and point to the promise of linking a public pedagogy of resistance that is university-based to broader struggles to change neoliberal policies. Equally important is that these movements connect learning to positive social change by making visible alternative models of radical democratic relations in a wide variety of sites, from the art gallery to alternative media to the university. Such movements offer instances of collective resistance to the glaring material inequities and the growing cynical belief that today's culture of investment and finance makes it impossible to address many of the major social problems facing both the United States and the world. These new forms of politics perform an important theoretical service by recognizing the important link among civic education, critical pedagogy, and oppositional politics. They also challenge the depoliticization of politics and provide new modes of resistance for promoting autonomy and democratic social transformation.

Student protesters against neoliberalism's assault on the university, public institutions, and civil society both understand how corporate capital works within various formations and sites—particularly the global media and the schools—and refuse to rely on dominant sources of information. These movements are developing an alternative form of politics outside of the party machines, a politics that astutely recognizes both the world of material inequality and the landscape of symbolic inequality.[98] In part, as we mentioned previously, this has resulted in what Imre Szeman calls "a new public space of pedagogy" that employs a variety of old and new media including computers, theater, digital video, magazines, the Internet, and photography as tools designed to link learning to social change, while creating networks that challenge the often hierarchical relations of more orthodox political organizations and cultural institutions.[99]

New forms of resistance have to be developed, demanding new forms of pedagogy and new sites in which to conduct it, while not abandoning traditional spheres of learning. The challenge for faculty in higher education is, in part, to find ways to contribute their knowledge and skills to understanding how neoliberalism devalues critical learning and undermines viable forms of political agency. Academics, as Imre Szeman puts it, need to figure out how neoliberalism and corporate culture "constitute a problem of and for pedagogy."[100] Academics need to be attentive to the oppositional pedagogies put into place by various student movements in order to judge their "significance . . . for the shape and function of the university curricula today,"[101] as well as their rhetorical and material impact on public spheres.

As we mentioned in chapter 3, faculty need to both support and learn as much as possible from student movements about establishing pedagogical approaches and political strategies that can be used to reclaim the university as a democratic public sphere. Faculty and students can work to reclaim higher education as a sphere where students learn not only about scholarly disciplines, diverse histories, and current theories, but also rethink the relationship among democracy, agency, and politics. In this scenario, education is not reduced to the acquisition of marketable skills, but provides the knowledge necessary for students to question unfettered power, exercise their role as critical and engaged citizens, and imagine a future in which the imperatives of an inclusive democracy rather than the demands of the national security state—with its suppression of dissent and civil liberties—become the organizing principle of everyday life, education, the nation, and the global public sphere.

We argue that any viable notion of higher education should be grounded in a vibrant politics and language that makes the promise of democracy a matter of concrete urgency. "Taking back higher education" means addressing the meaning and promise of democracy against its really existing forms, while understanding the vital role that education plays in making individual and collective actions possible. Taking back education represents both a referent for hope in a time of manufactured cynicism, fear, deception, and insecurity, and a call for action to be employed by an alliance of academics, students, activists, workers, and others who believe that the struggle over democracy is necessary as a check on injustice, the abuse of power, the depoliticization of the citizenry, and the corruption of education. The discourse of retaking education provides an ethical and political basis for both criticizing everywhere what parades as democracy—"the current state of all so-called democracy"[102]—and a language for critically assessing the conditions and possibilities for democratic transformation. Such a discourse embraces those values of an older republicanism in which civic courage, public service, and social responsibility reaffirm both the citizen as a critical agent and noncommodified public spheres as the sites from which the most important democratic values, identities, and social relations can be nurtured and experienced. We believe that the promise of democracy offers the proper articulation of a political ethics and suggests that when higher education is engaged through the project of democratic social transformation it can function as a vital public sphere for critical learning, ethical deliberation, and civic engagement.

Eric Gould argues that if the university is to provide a democratic education, it must "be an education for democracy . . . it must argue for its means as well as its ends . . . and participate in the democratic social process, displaying not only a moral preference for recognizing the rights of others and accepting them, too, but for encouraging argument and cultural critique." For Gould, higher education is a place for students to think critically and learn how to mediate between the imperatives of a "liberal democracy and the cultural contradictions of capitalism."[103] We think the university can do this and more. How might higher education become not just a place to think, but also a space in which to learn how to connect thinking with doing, critical thought with civic courage, knowledge with socially responsible action, citizenship with the obligations of an inclusive democracy? Knowledge must become the basis for considering individual and collective action, and it must reach beyond the university to join with other forces and create new public spheres in order to deal with the immense problems posed by neoliberalism and all those violations of human rights that negate the most basic premises of freedom, equality, democracy, and social justice. Higher education is also one of the few spheres in which freedom and privilege provide the conditions of possibility for teachers and students to act as critical intellectuals and address the inhumane effects of power, forge new solidarities across borders, identities, and differences, and also raise questions about what a democracy might look like that is inclusive, radically cosmopolitan and suited to the demands of a global public sphere.[104]

Under such circumstances, the meaning and purpose of higher education redefines the relationship between knowledge and power, on the one hand, and learning and social change on the other. Higher education as a democratic public sphere offers the conditions for resisting depoliticization, provides a language to challenge the politics of accommodation that subjects education to the logic of privatization, refuses to define students as simply consuming subjects, and actively opposes the view of teaching as a market-driven practice and learning as a form of training. At stake is not simply the future of higher education, but the nature of existing modes of democracy and the promise of an unrealized democracy—a democracy that promises a different future, one that is filled with hope and mediated by the reality of democratic-based struggles.[105]

Notes

Introduction: Why Taking Back Higher Education Matters

1. For some sources that have addressed this theme, see Jeffrey C. Goldfarb, *The Cynical Society: The Culture of Politics and the Politics of Culture in American Life* (Chicago: University of Chicago Press, 1991); Joseph N. Capella and Kathleen Hall Jamieson, *Spiral of Cynicism: The Press and the Public Good* (New York: Oxford University Press, 1997); Russell Jacoby, *The End of Utopia* (New York: Basic Books, 1999); William Chaloupka, *Everybody Knows: Cynicism in America* (Minneapolis: University of Minnesota Press, 1999); Zygmunt Bauman, *In Search of Politics* (Stanford: Stanford University Press, 1999); Carl Boggs, *The End of Politics: Corporate Power and the Decline of the Public Sphere* (New York: Guilford Press, 2000); Henry A. Giroux, *Public Spaces, Private Lives: Democracy Beyond 9-11* (Boulder, Colo.: Roman and Littlefield, 2003); Theda Skocpol, *Diminished Democracy* (Norman: University of Oklahoma Press, 2003).
2. On this issue, see Roberto Mangabeira Unger and Cornel West, *The Future of American Progressivism* (Boston: Beacon Press, 1998).
3. Cited in Bob Herbert, "Education Is No Protection," *The New York Times* (January 26, 2004), A27. Our italics.
4. Bob Herbert, "The Art of False Impression," *The New York Times* (August 11, 2003), A17.
5. Senator Robert Byrd, "From Bad to Worse . . . Billions for War on Iraq, a Fraction for Poor Kids Education," Senate floor remarks on September 5, 2003. Available on line at http://www.commondreams.org/views03/0906-09.htm.
6. W. E. B. Du Bois, *Against Racism: Unpublished Essays, Papers, Addresses, 1887–1961*, edited by Herbert Aptheker (Amherst: University of Massachusetts Press, 1985).
7. This issue is taken up by Stanley Aronowitz in *The Knowledge Factory* (Boston: Beacon Press, 2000).

Chapter 1 The Post-9/11 University and the Project of Democracy

1. Ulrich Beck, *The Reinvention of Politics: Rethinking Modernity in the Global Social Order*. Cambridge, UK: Polity Press, 1997, p. 78.

2. Testimony of Attorney General John Ashcroft to the Senate Committee on the Judiciary. December 6, 2001. Available at: http://www.usdoj.gov/ag/testimony/2001/1206transcriptsenatejudiciarycommittee.htm.

3. This essay was written by Susan Searls Giroux; an earlier version of it appeared in JAC.

4. I'm borrowing the concept of "democratization" over "democracy" as the aim of progressive politics from Samir Amin in "Imperialism and Globalization," *Monthly Review* (June 2001), 6–24.

5. Douglas Kellner, "Reading the Gulf War: Production, Text, Reception," in *Media Culture: Cultural Studies, Identity, and Politics between the Modern and the Postmodern* (New York: Routledge, 1995), 199.

6. *Ibid.*, 210.

7. *Ibid.*, 218.

8. *Ibid.*, 217.

9. *Ibid.*, in *The Castoriadis Reader*, ed. David Ames Curtis (Malden, Mass: Blackwell, 1997), 220.

10. *Ibid.*

11. Castoriadis, Cornelius. "Done and To Be Done." *The Castoriadis Reader.* Trans. and Ed. David Ames Curtis (Oxford, Eng.: Blackwell, 1997), 400.

12. Zygmunt Bauman, *In Search of Politics* (Stanford: Stanford University Press, 1999), 85.

13. *Ibid.*, 86.

14. Ulrick Beck, *The Reinvention of Politics: Rethinking Modernity in the Global Social Order*, trans. Mark Ritter (Cambridge, Mass: Blackwell, 1997), 163.

15. Nick Couldry, "A Way Out of the (Televised) Endgame?" *Open Democracy.* http://www.opendemocracy.net/forum/document_details.asp?CatID=5&DocID=712.p.1.

16. Ari Fleischer, cited in Purdum, Todd S., and Alison Mitchell. "Bush, Angered by Leaks, Duels with Congress." *The New York Times* (October 10, 2001), A1[+].

17. Elizabeth Becker, "In the War on Terrorism, a Battle to Shape Public Opinion." *The New York Times* (November 11, 2001), 11.

18. Couldry, 2.

19. Jerry L. Martin and Anne D. Neal, *Defending Civilization: How Our Universities are Failing America and What Can Be Done About It.* http://www.goacta.org/reportsframeset.htm.1.

20. *Ibid.*, 1.

21. *Ibid.*, 3.

22. *Ibid.*, 8.

23. *Ibid.*, Curtis White, "The New Censorship," *Harper's Magazine* (August 2003), 16.

24. *Ibid.*, 7. Paul Street, in an article entitled, "By All Means, Study the Founders: Notes from the Democratic Left" elaborates in brilliant detail the ironies of this position. See his article in *The Review of Education/Pedagogy/Cultural Studies* 25: 4, forthcoming, Winter 2003.

25. *Ibid.*, 7.

26. Cited by Martin and Neal, *Ibid.*, 7.

27. Lynne Cheney, *Telling the Truth: Why Our Culture and Our Country Stopped Making Sense and What We Can Do About It* (New York: Simon and Schuster, 1995), 1.

28. *Ibid.*, 1.

29. Diane Ravitch, "Now is the Time to Teach Democracy," *Education Week* (October 17, 2001), 48.
30. *Ibid.*, 48.
31. Judith Butler, "Explanation and Exoneration, or What We Can Hear," *Theory and Event* 5: 4 at http://muse.jhe.edu/journals/theory_and_event/v005/5/4butler.html.p.21.
32. Edward Rothstein, "Attacks on the U.S. Challenge the Perspectives of Postmodern True Believers" *New York Times* (September 22, 2001), A17.
33. *Ibid.*
34. Waller R. Newell, "Postmodern Jihad: What Osama bin Laden Learned from the Left," *The Weekly Standard* Vol. 007. Issue 11 (November 26, 2001). Available at: http://www.weeklystandard.com/content/public/articles/000/000/000/553fragu.asp.
35. *Ibid.*
36. Daryn Kagan, Host, "People in the News: The American Taliban: John Walker." Aired December 22, 2001. Transcript available at: http://www.cnn.com/TRANSCRIPTS/0112/22/pitn.00.html.
37. *Ibid.*
38. *Ibid.*
39. Jay Bookman, "Ann Coulter Wants to Execute You," *The Atlanta Journal-Constitution* (February 14, 2002), available on line: http://www.indybay.org/news/2002/02/116560.php
40. Todd Gitlin, "Blaming America First," *Mother Jones* (January/February 2002), 22.
41. *Ibid.*
42. *Ibid.*, 24.
43. Robert McChesney, "Introduction" in Noam Chomsky, *Profit Over People* (New York: Seven Stories Press, 1999), 7.
44. *Ibid.*, 9.
45. Bauman, *In Search of Politics*, 65–66.
46. Beck, *The Reinvention of Politics*, 83.
47. Castoriadis, "Democracy as Procedure and Democracy as Regime," *Constellations* 4 (1997), 11.
48. *Ibid.*
49. Jacques Derrida, "Mochlos; or the Conflict of the Faculties," in *Logomachia: The Conflict of the Faculties*, ed. Richard Rand (Lincoln: University of Nebraska Press, 1992), 3.
50. Cornelius Castoriadis, "The Greek *Polis* and the Creation of Democracy." *Philosophy, Politics, Autonomy: Essays in Political Philosophy*. Ed. David Ames Curtis (New York: Oxford University Press, 1991), 113.
51. *Ibid.*
52. Cornelius Castoriadis, "Democracy," 10.
53. Bauman, *In Search of Politics*, 86.
54. *Ibid.*
55. Curtis, "Translator's Foreword" in Cornelius Castoriadis. *The World in Fragments*, ed. and trans. by David Ames Curtis (Stanford: Stanford University Press, 1997), xi.
56. Bauman, *In Search of Politics*, 87.
57. *Ibid.*
58. *Ibid.*, 65.
59. *Ibid.*, 107.

60. Cornelius Castoriadis, "Done and To Be Done," 407.

61. Bauman, *In Search of Politics*, 8.

62. See, for example, Ian Hunter's *Culture and Government: The Emergence of Literary Education* (London: MacMillan, 1998), Tony Bennett's *Outside Literature* (London and New York: Routledge, 1990) and his more recent *Culture: A Reformer's Science* (London: Sage, 1998) for a notion of pedagogy that functions more or less exclusively within the logic of governmentality.

63. Bauman, *In Search of Politics*, 99. Bill Readings has made a similar argument. See his *The University in Ruins* (Cambridge, Mass.: Harvard University Press, 1996).

64. See, for example, Bauman's *In Search of Politics* (Stanford: Stanford University Press, 1999), Henry Giroux's *Public Spaces, Private Lives: Beyond the Culture of Cynicism* (Lanham: Rowman and Littlefield, 2001), and Carl Boggs's *The End of Politics: Politics, Corporate Power and the Decline of the Public Sphere* (New York: Gulford, 2000).

65. See Boggs's *The End of Politics*.

66. See Stanley Aronowitz, *The Knowledge Factory: Dismantling the Corporate University and Creating True Higher Learning* (Boston: Beacon Press, 2000).

67. Les Terry, "Traveling 'The Hard Road to Renewal': A Continuing Conversation with Stuart Hall," *Arena Journal* 8 (1997), 47.

68. See Pierre Bourdieu, "For a Scholarship with Commitment," *Profession* (2000), 40–45.

69. Beck, *The Reinvention of Politics*, 83.

70. See Richard Rorty, "The Unpatriotic Academy," originally published in *The New York Times* op ed section. The essay is reprinted in his *Achieving Our Country: Leftist Thought in the Twentieth Century* (Cambridge, Mass.: Harvard University Press, 1998).

71. Raymond Williams, *The Politics of Modernism: Against the New Conformists* (New York: Verso, 1989), 175.

72. Burton Bledstein, *The Culture of Professionalism: The Middle Class and the Development of Higher Education in America* (New York: Norton, 1976), 327–28.

73. Bauman, *In Search of Politics*, 74.

74. Raymond Williams, *Communications*, rev. ed. (New York, Barnes & Noble, 1966), 15–16. Emphasis added.

75. *Ibid.*, 15.

76. *Ibid.*, 16.

Chapter 2   Academic Culture, Intellectual Courage, and the Crisis of Politics in an Era of Permanent War

1. Pierre Bourdieu, "For a Scholarship with Commitment," *Profession 2000* (2000), 40.

2. Carol A. Stabile and Junya Morooka, " 'Between Two Evils, I Refuse to Choose the Lesser,' " *Cultural Studies* 17: 3 (2003), 326.

3. Bourdieu, 41.

4. Pierre Bourdieu et al., *The Weight of the World*, trans. Priscilla Parkhust Ferguson et al. (Stanford, Cal.: Stanford University Press, 1999, 1st ed. 1993), 629.

5. Bourdieu, cited in Andrea Noll, "Counter-Fire—in Memoriam Pierre Bourdieu, July 20, 2003," *ZNet Commentary*. Available on line: http://www.zmag.org/ sustainers/content/2003-07/15noll.cfm.

6. Cited by Bill Moyers on *NOW*, aired on May 14, 2003.

7. Cited in Robert Dreyfuss, "Grover Norquist: 'Field Marshal' of the Bush Plan," *The Nation* (May 14, 2001).

8. Slavo Žižek, "Today Iraq. Tomorrow. . . . Democracy?" *In These Times* (May 5, 2003), 28.

9. George Steinmetz, "The State of Emergency and the Revival of American Imperialism; Toward an Authoritarian Post-Fordism," *Public Culture* 13:2 (Spring 2003), 329.

10. Dick Cheney cited in Sam Dillon, "Reflections on War, Peace, and How to Live Vitally and Act Globally," *The New York Times* (June 1, 2003), 28.

11. Fort Stewart passage: cited in "Coming Home," *Now with Bill Moyers* (November 11, 2003). Available on line: www.pbs.org≠now≠archive_ transcripts.html. Bob Herbert, "Oblivious in D.C.," *The New York Times* (June 30, 2003).

12. Bill Moyers, "Deep in a Black Hole of Red Ink," Common Dreams News Center (May 30, 2003). Available on line at: www.commondreams.org/ views03/0530–11.htm.

13. Cited in Editorial Comment, "Bush's Sleight of Hand," *The Progressive* (July 2003), 8.

14. Zygmunt Bauman, *Society Under Siege* (Malden, Mass.: Blackwell, 2002), 54.

15. Arundhati Roy, *War Talk* (Boston: South End Press, 2003), 34. Bush and God passage: cited in Paul Harris, "Bush Says God Chose Him to Lead His Nation," *The Observer* (November 1, 2003).

16. President George W. Bush, address to joint session of Congress, "September 11, 2001, Terrorist Attacks on the United States." Available on line: www.cnn.com/ 2001/US/09/11/bush.speech.text. Bush repeated this position in one of his public ads for the 2004 presidential campaign.

17. Sheldon Wolin, "Inverted Totalitarianism: How the Bush Regime is effecting the Transformation to a Fascist-Like State," *The Nation* (May 19, 2003), 13.

18. Ulrich Beck, "The Silence of Words and Political Dynamics in the World Risk Society," *Logos* 1: 4 (Fall 2002), 1.

19. Susan Buck-Morss, *Thinking Past Terror: Islamism and Critical Theory on the Left* (London: Verso Press, 2003), 30–31.

20. Beck, 3.

21. Jeff Madrick, "The Iraqi Time Bomb," *The New York Times Magazine* (April 6, 2003), 50.

22. These ideas are taken from Benjamin R. Barber, "Blood Brothers, Consumers, or Citizens? Three Models of Identity—Ethnic, Commercial, and Civic," in Carol Gould and Pasquale Pasquino, eds., *Cultural Identity and the Nation State* (Lanham, Md.: Rowman and Littlefield, 2001), 65.

23. Edward S. Herman and Robert W. McChesney, *The Global Media: The New Missionaries of Global Capitalism* (Washington and London: Cassell, 1997), 3.

24. I address this issue in Henry A. Giroux, *Public Spaces, Private Lives: Democracy Beyond 9/11* (Lanham, Md.: Rowman and Littlefield, 2003).

25. Barber, 59.

26. Editorial, "Capitalism and Democracy," *The Economist* (June 28–July 4, 2003), 13.

27. Zygmunt Bauman provides the same sort of analysis in analyzing the state of the poor under neoliberalism. He argues the poor are excluded and often charged with the guilt of their exclusion, and thus seen as dangerous and a threat to society. See Zygmunt Bauman, *Work, Consumerism and the New Poor* (Philadelphia: Open University Press, 1998), especially chapter 5, "Prospects for the Poor," 83–98.

28. Cited in Mike Davis, "Cry California," *Common Dreams News Center* (September 4, 2003). Available on line at: http://www.commondreams. org/views03/0903-10.htm.

29. See Bauman, *Work, Consumerism and the New Poor,* 93.

30. U.S. Senator Robert Byrd, "Reckless Administration May Reap Disastrous Consequences," Senate Floor Speech, February 12, 2003. Available on line at: http://www.commondreams.org/views03/0212–07.htm.

31. Published by the BBC, "Child Sickness 'Soars' in Iraq," June 9, 2003. Available on line at: http://www.comondreams.org/headlines03/0609–05.htm.

32. John Pilger, "Who Are the Extremists?" Znet (August 22, 2003). Available on line: http://www.zmag.org/sustainers/content/2003–08/23pilger.cfm.

33. Milton Friedman, *Capitalism and Freedom* (Chicago: University of Chicago Press, 2002 reprint), 12.

34. Ibid., 33.

35. Steven R. Donziger, ed. *The Real War on Crime: The Report of the National Criminal Justice Commission* (New York: Harper Perennial, 1996), 101.

36. Paul Street, "Race, Prison, and Poverty: The Race to Incarcerate in the Age of Correctional Keynesianism," *Z Magazine* (May 2001), 25.

37. Peter Marcuse is right to argue that the state under neoliberal globalization does not lack power; it abdicates state power. See Peter Marcuse, "The Language of Globalization," Monthly Review 52: 3 (July–August 2000). Available on line at: http://www.monthlyreview.org/700marc.htm.

38. Marian Wright Edelman, "Standing Up for the World's Children: Leave No Child Behind," Available on line at: http://gos.sbc.edu/e/edelman.html.

39. Jane Sutton-Redner, "Children in a World of Violence," *Children In Need Magazine* (December 2002). Available on line at http://childreninneed.com/magazine/violence/html.

40. Robert W. McChesney, "Global Media, Neoliberalism, and Imperialism," *Monthly Review* (March 2001), 16.

41. Manuel Castells, *End of Millennium, III* (Malden, Mass.: Blackwell, 1998), 149. Of course, while Castells is right about the conditions of labor, he under emphasizes the huge surplus of labor around the world, This is made clear in another International Labour Organization (ILO) report that states that about a quarter of the labor force around the world is unemployed and a third under-employed.

42. This information from the ILO is available on line at http://us.ilo.org/ilokidsnew/ILOU/101.html.

43. "End Child Prostitution, Child Pornography, and the Trafficking of Children for Sexual Exploitation (ECPAT)", *Europe and North America Regional Profile*, issued by the World Congress Against Commercial Sexual Exploitation of Children (held in Stockholm, Sweden, August 1996), 70.

44. For an extensive record of such treaty and protocol refusals on the part of the Bush administration, see Richard Du Buff, "Mirror Mirror on the Wall, Who's

the Biggest Rogue of All?," *ZNet* (August 7, 2003). Available on line at: Znetupdates@Zmail.zmag.org. pp. 1–5.

45. Buck-Morss, 10.

46. Lynn Worsham and Gary A. Olson, "Rethinking Political Community: Chantal Mouffe's Liberal Socialism," *Journal of Composition Theory* 19: 2 (1999), 178.

47. Samir Amin, "Imperialization and Globalization," *Monthly Review* (June 2001), 16.

48. Bob Herbert, "The Money Magnet," *The New York Times* (June 23, 2003), A27.

49. Fathali M. Moghaddam, "Health, Beauty, and Seniors," *Health and Age* (July 4, 2003). Available on line: http://www.healthandage.com/Home/gm=20! gid2=2515!gnews=01030703.

50. The wonderful distinction between "hegemony lite" and hegemony from which we have developed this idea comes from Kate Crehan, *Gramsci, Culture, and Anthropology* (Los Angeles: University of California Press, 2002), 202–07.

51. This theme is developed with great care in Martha C. Nussbaum, "Compassion & Terror," *Daedalus* (Winter 2003), 10–26, especially 12.

52. Buck-Morss, 103–04.

53. Edward Said, "On Defiance and Taking Positions," *Reflections on Exile and Other Essays* (Cambridge, Mass.: Harvard University Press, 2001), 501.

54. Stabile and Morooka, 338.

55. See Arundhati Roy, *Power Politics* (Cambridge, Mass.: South End Press, 2001).

56. Pierre Bourdieu and Gunter Grass, "The Progressive Restoration: A Franco-German Dialogue," *New Left Review* 14 (March–April, 2002), 63–67.

57. Alain Touraine, *Beyond Neoliberalism* (London: Polity, 2001), 50.

58. *Ibid.*, 33.

59. George Merritt and John Ingold, "Republicans Seek 'Balance' on College Faculties," *The Denver Post* (September 8, 2003). Available on line at: http://www.denverpost.com/stories/0,1413,36%257e53%257e1617905,00.html.

Chapter 3   Cultural Studies and Critical Pedagogy in the Academy

1. Raymond Williams, "The Future of Cultural Studies," in *The Politics of Modernism* (London: Verso, 1989), 161–62.

2. Alain Badiou, *Ethics: An Essay on the Understanding of Evil* (London: Verso, 1998), 11.

3. Jonathau Rowe, "Wasted Work, Wasted Time," in John De Graf. ed., *Take Back Your Time* (San Francisco, Cal.: Berrett-Koehler Publishers, 2003), 60.

4. Nick Couldry, "In the Place of a Common Culture, What?" *Review of Education/Pedagogy/Cultural Studies* 26:1 (2004), p. 10.

5. For instance, in a number of readers on cultural studies, the issue of critical pedagogy is left out altogether. Typical examples include: Toby Miller, ed., *A Companion to Cultural Studies* (Malden, Mass.: Blackwell, 2001); Simon During, ed., *The Cultural Studies Reader, Second Edition* (New York: Routledge, 1999); and John Storey, ed., *What is Cultural Studies?: A Reader* (New York: Arnold Press, 1996).

6. Shane Gunster, "Gramsci, Organic Intellectuals, and Cultural Studies," in Jason A. Frank and John Tambornino, eds., *Vocations of Political Theory* (Minneapolis: University of Minnesota Press, 2000), 253.

7. See for example Roger Simon, *Teaching Against the Grain: Texts for a Pedagogy of Possibility* (Westport, Conn.: Bergin and Garvey, 1992); Henry A. Giroux, *Border Crossings: Cultural Workers and the Politics of Education* (New York: Routledge, 1992); David Trend, *Cultural Pedagogy: Art/Education/Politics* (Westport, Conn.: Bergin and Garvey, 1992).

8. Richard Johnson, "Reinventing Cultural Studies: Remembering for the Best Version," in Elizabeth Long, ed., *From Sociology to Cultural Studies* (Malden, Mass.: Basil Blackwell, 1997), 452–88. See also the brilliant early essay by Grossberg on education and cultural studies in Lawrence Grossberg, "Introduction: Bringin' It all Back Home—Pedagogy and Cultural Studies," in Henry A. Giroux and Peter McLaren, eds., *Border Crossings: Pedagogy and the Politics of Cultural Studies* (New York: Routledge, 1994), 1–25.

9. I take this issue up in greater detail in Henry A. Giroux, *Impure Acts: The Practical Politics of Cultural Studies* (New York: Routledge, 2000); *Stealing Innocence: Corporate Culture's War on Children* (New York: Palgrave, 2000).

10. Roger Simon, "Broadening the Vision of University-Based Study of Education: The Contribution of Cultural Studies," in *The Review of Education/Pedagogy/Cultural Studies* 12: 1 (1995), 109.

11. On the importance of problematizing and pluralizing the political, see Jodi Dean, "The Interface of Political Theory and Cultural Studies," in Jodi Dean, ed., *Cultural Studies and Political Theory* (Ithaca, NY: Cornell University Press, 2000), 1–19.

12. Williams, *Communications*, 14.

13. See especially Raymond Williams, *Marxism and Literature* (New York: Oxford University Press, 1977); and Raymond Williams, *The Year 2000* (New York: Pantheon, 1983).

14. Williams, *Marxism and Literature, Ibid.*

15. Matthias Fritsch, "Derrida's Democracy to Come," *Constellations* 9: 4 (December 2002), 579.

16. Lawrence Grossberg, "The Circulation of Cultural Studies," in Storey, *What is Cultural Studies?* 179–80.

17. Imre Szeman, "Learning to Learn from Seattle," *The Review of Education/Pedagogy/Cultural Studies* 24: 1–2 (2002), 4.

18. Lawrence Grossberg, "Toward a Genealogy of the State of Cultural Studies," in Cary Nelson and Dilip Parameshwar Gaonkar, eds., *Disciplinarity and Dissent in Cultural Studies* (New York: Routledge, 1996), 143.

19. Douglas M. Kellner and Meenakshi Gigi Durham, "Adventures in Media and Cultural Studies: Introducing Key Works," in Douglas M. Kellner and Meenakshi Gigi Durham, eds., *Media and Cultural Studies: Key Works* (Malden, Mass.: Blackwell, 2001), 29.

20. Henry A. Giroux, *The Mouse that Roared: Disney and the End of Innocence* (Lanham, Md.: Rowman and Littlefield, 2000); see also Henry A. Giroux, *Impure Acts* (New York: Routledge, 2000).

21. Henry A. Giroux, *Breaking into the Movies: Film and the Politics of Culture* (Malden, Mass.: Blackwell, 2002).

22. See, for instance, Durham and Kellner, *Media and Cultural Studies*. The concept of sites of pedagogy and pedagogical address can be found in the work of

Jeffrey DiLeo. See, for example, Jeffrey DiLeo et al., "The Sites of Pedagogy," *Sympolke* 10: 1 (2002), 7–12.

23. On this issue, see Danny Goldberg, *Dispatches from the Culture Wars: How the Left Lost Teen Spirit* (New York: Miramax Books, 2003).

24. James Clifford, "Museums in the Borderlands," in *Different Voices*, ed. Carol Becker et al. (New York: Association of Art Museum Directors, 1992), 129.

25. James Young, *At Memory's Edge* (New Haven, Conn.: Yale University Press, 2000), 182.

26. John Beverly, "Pedagogy and Subalternity: Mapping the Limits of Academic Knowledge," in Rolland G. Paulston, ed., *Social Cartography* (New York: Garland, 1996), 354.

27. See Henry A. Giroux and Peter McLaren, eds., *Between Borders: Pedagogy and the Politics of Cultural Studies* (New York: Routledge, 1993).

28. Amy Gutmann, *Democratic Education* (Princeton, NJ: Princeton University Press, 1998), 42.

29. Ien Ang, "On the Politics of Empirical Audience Research," in Durham and Kellner, *Media and Cultural Studies*, 183.

30. Arif Dirlik, "Literature/Identity: Transnationalism, Narrative, and Representation," *Review of Education/Pedagogy/Cultural Studies* 24: 3 (July–September 2002), 218. One example of such work can be found in Marjorie Garber, *Sex and Real Estate: Why We Love Houses* (New York: Anchor Books, 2000) or even better Marjorie Garber, *Dog Love* (New York: Touchstone Press, 1997).

31. Stuart Hall cited in Peter Osborne and Lynne Segal, "Culture and Power: Stuart Hall Interviewed," *Radical Philosophy* N. 86 (November/December 1997), 24.

32. See, for example, Douglas Kellner, *Media Culture: Cultural Studies, Identity, and Politics* (New York: Routledge, 1995).

33. Zygmunt Bauman, *Society under Siege* (Malden, Mass.: Blackwell: 2002), 170.

34. Some recent examples include: Karen Kopelson, "Rhetoric on the Edge of Cunning; or, The Performance of Neutrality (Re)Considered as a Composition Pedagogy for Student Resistance," CCC (September 2003), 115–46; Allyson P. Polsky, "Argument, Evidence, and Engagement: Training Students as Critical Investigators and Interpreters of Rhetoric and Culture," *Pedagogy* 3: 3 (2003), 427–30; Gerald Graff, *Beyond the Culture Wars: How Teaching the Conflicts can Revitalize American Education* (New York: W. W. Norton, 1993).

35. Gerald Graff appears to have made a career out of this issue by misrepresenting the work of Paulo Freire and others; citing theoretical work by critical educators that is outdated and could be corrected by reading anything they might have written in the last five years; creating caricatures of their work; or by holding up as an example of what people in critical pedagogy do (or more generally anyone who links pedagogy and politics) the most extreme and ludicrous examples. For more recent representations of this position, see Gerald Graff, "Teaching Politically Without Political Correctness," *Radical Teacher* 58 (Fall 2000), 26–30; Gerald Graff, *Clueless in Academe* (New Haven, Conn.: Yale University Press, 2003).

36. Lani Guinier, "Democracy Tested," *The Nation* (May 5, 2003), 6. Guinier's position is in direct opposition to that of Graff and his acolytes. For instance, see "A Conversation Between Lani Guinier and Anna Deavere Smith, 'Rethinking Power, Rethinking Theater,'" *Theater* 31: 3 (Winter 2002), 31–45.

37. George Lipsitz, "Academic Politics and Social Change," in Dean, *Cultural Studies and Political Theory*, 81–82.
38. Steve Benton, "Concealed Commitment," *Pedagogy* 1: 2 (2001), 254.
39. For a more detailed response to this kind of watered-down pedagogical practice, see Stanley Aronowitz, *The Knowledge Factory* (Boston: Beacon Press, 2000); Henry A. Giroux, *The Abandoned Generation: Democracy Beyond the Culture of Fear* (New York: Palgrave, 2003).
40. One of the most egregious, if not silly, examples of this kind of work can be found in *Education Next*, a magazine edited by conservative hardliners Paul Peterson and Chester Finn, Jr. See J. Martin Rochester, "Critical Demagogues," *Education Next* 3: 4 (2003), 77–82. Rochester is so intent on arguing that critical pedagogy is ideological—as if its advocates do not admit that they work out of particular ideologies and political projects—that he completely ignores his own ideologically driven agenda (again pushing the old and stale claim by the right that they are ideologically neutral!). The result is an utter lack of critical self-consciousness and endless series of foolish assertions parading as arguments. The only saving feature of the piece is that its misrepresentations and distortions read like a parody rather than a serious engagement with critical pedagogy, and ultimately provide fodder for comic relief. Rochester is so clueless that he actually argues that Henry Giroux and Peter McLaren produce identical work. If Rochester had spent 30 minutes examining the books of these authors, especially those written after 1995, he would have recognized that McLaren works out of a Marxist framework while Giroux's work is indebted to the tradition of radical democracy. It gets worse. The article lacks any understanding of the scholarship produced by critical pedagogy theorists, has no sense of the complexities, differences, and history of the field and, ultimately, is reduced to a morality play replete with the simplified scenario of good versus evil that has become standard fare for this kind of right-wing hatchet job parading as serious scholarship.
41. An interview with Julie Ellison, "New Public Scholarship in the Arts and Humanities," *Higher Education Exchange* (2002), 20.
42. Amy Gutmann, *Democratic Education* (Princeton, NJ: Princeton University Press, 1998), 42.
43. Cited in Emily Eakin, "The Latest Theory is that Theory Doesn't Matter," *The New York Times* (April 19, 2003), A17.
44. Stanley Fish, "Aim Low," *The Chronicle of Higher Education* (May 16, 2003). Available on line: http://chronicle.com/jobs/2003/05/2003051601c.htm.
45. *Ibid.*
46. Edward W. Said, "The Public Role of Writers and Intellectuals," *The Nation* (October 1, 2001), 31.
47. The entire quote is: "We think we've been through a period where too many people have been given to understand that if they have a problem, it's the government's job to cope with it. 'I have a problem, I'll get a grant.' 'I'm homeless, the government must house me.' They're casting their problem on society. And, you know, there is no such thing as society. There are individual men and women, and there are families. And no government can do anything except through people, and people must look to themselves first. It's our duty to look after ourselves and then, also to look after our neighbour. People have got the entitlements too much in mind, without the obligations. There's no such thing

as entitlement, unless someone has first met an obligation." Margaret Thatcher, "Aids, Education, and the Year 2000," *Women's Own Magazine* (October 3, 1987), 10.

48. Geoffrey Hartman, "Public Memory and Its Discontents," *Raritan* 8: 4 (Spring 1994), 28.

49. One might be tempted to view the arguments posed by Graff and Fish as an idiosyncracy safely confined to the ivory tower of the University of Illinois at Chicago. The presumption here is that one can aspire either to high academic achievement or fall prey to the latest academic fashions such as multicultural-ism or critical pedagogy. Such views have been echoed by the conservative National Association of Scholars, Lynne Cheney, William Bennett, and others both within and outside of higher education, and most recently can be seen in the national debates over affirmative action in which rigor is pitted against diversity, as if one can have one without the other.

50. The full quote is worth repeating: "On more than one occasion I have had an experience many of you will recognize. A student you haven't seen in years rushes up to you and says, 'Oh, Professor, I think so often of that class in 1995 (or was it 1885) when you said X and I was led by what you said to see Y and began on that very day to travel the path that has now taken me to success in profession Z. I can't thank you enough.' You, however, are appalled, because you can't imagine yourself ever saying X (in fact you remember spending the entire semester saying anti-X) and you would never want anyone to exit from your class having learned Y (A lesson you have been preaching against for 20 years) and you believe that everyone would be better off if profession Z dis-appeared from the face of the earth. What, you might ask, did I do wrong?" Stanley Fish, "Aim Low," *The Chronicle of Higher Education* (May 16, 2003). Available on line: http://chronicle.com/jobs/2003/05/2003051601c.htm. Actually, as my friend, Michael Brenson, mentioned to me, we would prefer many forms of sentimentality to Fish's smug rationalism and stark distrust of the personal.

51. Some academics see Fish as being on the left. I think Terry Eagleton rightly con-textualizes this view with his comment: "It is one of the minor symptoms of the mental decline of the United States that Stanley Fish is thought to be on the Left." See Terry Eagleton, "Stanley Fish," in *Figures of Dissent* (London: Verso, 2003), 171.

52. On students being pushed out of public schools, see the excellent articles by Tamar Lewin and Jennifer Medina, "To Cut Failure Rate, Schools Shed Students," *The New York Times* (July 31, 2003), A1, A22 and "High School Under Scrutiny for Giving Up on Its Students," *The New York Times* (August 1, 2003) A1, A22. For one of the most notorious examples of schools covering up their dropout rates, see Michael Winerip, "A 'Zero Dropout' Miracle: Alas! A Texas Tall Tale," *The New York Times* (August 13, 2003), A19.

53. Information on The Free Child project can be found on http://freechild.org. On the work of the Urban Debate League, see Dana Williams, *Urban Debate Leagues* (September 3, 2003). Available on line: http://www.tolerance.org/teens/stories/article.jsp?p=0&ar=58. pp. 1–5.

54. See, for example, the recent work of Theda Skocpol, *Diminished Democracy: From Membership to Management in American Civic Life* (Norman, Okla.: University of Oklahoma, 2003).

55. Arundhati Roy, *Power Politics* (Cambridge, Mass.: South End Press, 2001), 6.

56. Edward N. Said, *Reflections on Exile and Other Essays* (Cambridge, Mass.: Harvard University Press), 503.
57. Susan Buck-Morss, *Thinking Past Terror: Islamism and Critical Theory on the Left* (London: Verso Press, 2003), 21.
58. Bill Readings, *The University in Ruins* (Cambridge, Mass.: Harvard University Press), 11, 18.
59. Zygmunt Bauman, *In Search of Politics* (Stanford: Stanford University Press, 1999), 170.
60. Zygmunt Bauman, *Society under Siege* (Malden, Mass.: Blackwell, 2002), 70.
61. *Ibid.*
62. Lynn Worsham and Gary A. Olson, "Rethinking Political Community: Chantal Mouffe's Liberal Socialism," *Journal of Composition Theory* 19: 2 (1999), 178.
63. Jo Ellen Green Kaiser, "A Politics of Time and Space," *Tikkun* (November/December 2003), 17–18.
64. For a detailed analysis of linking the crisis of time to the crisis of democracy, see John de Graff, ed., *Take Back Your Time: Fighting Overwork and Time Poverty in America* (San Francisco: Berrett-Koehler Publishers, Inc., 2003).
65. Alain Badiou, *Ethics: An Essay on the Understanding of Evil* (London: Verso, 1998), 115–16.
66. Dick Hebdige, "Training Some Thoughts on the Future," in Jon Bird et al., eds., *Mapping the Futures* (New York: Routledge, 1993), 275.
67. Noam Chomsky, "Paths Taken, Tasks Ahead," *Profession* (2000), 34.
68. Pierre Bourdieu, "For a Scholarship With Commitment," *Profession* (2000), 44.
69. Roy, *Power Politics*, 12.
70. Guinier and Smith, 34–35.

Chapter 4   Race, Rhetoric, and the Contest over Civic Education

1. See for example *Race, Rhetoric, and the Postcolonial* (1999), ed., Gary A. Olson and Lynn Worsham; *Pedagogy of Freedom: Ethics, Democracy, and Civic Courage* (1998), by Paulo Freire; and *Race, Rhetoric, and Composition* (1999), ed. Keith Gilyard.
2. Stanley Aronowitz, "Introduction" to Freire, *Pedagogy of Freedom: Ethics, Democracy, and Civic Courage* (Lanham: Rowman and Littlefield, 1998), 6.
3. Quoted in Eyal Press and Jennifer Washburn, "The Kept University," *The Atlantic Monthly* (March 2000), 52.
4. *Ibid.*
5. See also Stanley Aronowitz, *The Knowledge Factory: Dismantling the Corporate University and Creating True Higher Learning* (Boston: Beacon Press, 2000), for a provocative assessment of contemporary university life as well as a proposal for a new curriculum that takes seriously the reform strategies offered by both right and left.
6. Judith N. Shklar, *American Citizenship: The Quest for Inclusion* (Cambridge, Mass.: Harvard University Press, 1991), 1.
7. Quoted in Gary Olson and Lynn Worsham, eds., *Race, Rhetoric and the Postcolonial* (Albany: State University of New York Press, 1999), 182.

8. I borrow these categories for distinguishing among different versions of citizenship from Rogers M. Smith, *Civic Ideals: Conflicting Visions of Citizenship in U.S. History* (New Haven, Conn.: Yale University Press, 1997).

9. Stuart Hall, "Variants on Liberalism" in *Politics and Ideology*, ed. Stuart Hall and James Donald (Milton Keynes: Open University Press, 1986), 39.

10. *Ibid.*

11. For a further elaboration of liberalism's core principles, as well as a critical assessment of its response to racism, see David Theo Goldberg, *Racist Culture: Philosophy and the Politics of Meaning* (Cambridge: Blackwell, 1993). See also Stuart Hall, "Variants on Liberalism," 34–69.

12. As Judith Shklar has convincingly argued, citizenship is as much about the right to vote as the right to achieve independent social status, or standing—in short, to reap the benefits of one's own labor. Those who did not own their work (women and slaves) thus lacked the capacities for citizenship.

13. Philip Gleason, "American Identity and Americanization" in William Peterson, Michael Novak, and Philip Gleason, *Concepts of Ethnicity* (Cambridge, Mass.: Belknap, 1980), 62–63.

14. This view can also be found in: H. Kohn's *American Nationalism* (1957), Michael Harrington's *Decade of Decision: The Crisis of the American System* (1980), and Samuel Huntington's *American Politics: The Promise of Disharmony* (1981).

15. Rogers M. Smith, "The 'American Creed' and American Identity: The Limits of Liberal Citizenship in the United States," *Western Political Quarterly* 41 (1988), 230–31.

16. Chantal Mouffe, *The Return of the Political* (New York: Verso, 1993), 62.

17. Adrian Orfield, "Citizenship and Community: Civic Republicanism and the Modern World," in *Citizenship Debates: A Reader* (Minneapolis: University of Minnesota Press, 1998), 79.

18. Quoted in Rogers Smith, "The American Creed," 231.

19. *Ibid.*

20. *Ibid.*

21. Shklar, *American Citizenship*, 1.

22. Rogers M. Smith, *Civic Ideals: Conflicting Visions of Citizenship in U.S. History* (New Haven, Conn.: Yale University Press, 1997), 2–3.

23. Smith, "The American Creed," 234.

24. See, for example, Lawrence Cremin, *American Education: The Metropolitan Experience 1870–1980* (New York: Harper, 1988).

25. Matthew Frye Jacobson, *Whiteness of a Different Color* (Cambridge, Mass.: Harvard University Press, 1998), 23.

26. *Ibid.*, 26.

27. *Ibid.*

28. *Ibid.*, 27.

29. David Theo Goldberg, *Racist Culture: Philosophy and the Politics of Meaning* (London: Blackwell, 1994), 6–7.

30. See J. Higham, *Strangers in the Land: Patterns of American Nativism, 1860–1925* (New York: Atheneum, 1966/1955); Reginald Horsman, *Race and Manifest Destiny* (Cambridge, Mass.: Harvard University Press, 1981); George Frederickson, *White Supremacy: A Comparative Study in American and South*

African History (Oxford: Oxford University Press, 1981); and David Theo Goldberg, *Racist Culture*.

31. Ivan Hannaford, *Race: The History of an Idea in the West* (Baltimore: Johns Hopkins University Press, 1996), 10–11, emphasis added. In mapping the decline of politics and the rise of the racial state, Hannaford also details the history of European intellectual misreadings of classical texts, particularly the projection of modern racial classifications onto antiquity. Additionally, Hannaford comments on the shift in European philosophy away from Aristotle's *Politics* and an ever-growing interest in his *Poetics*, a point of interest in our efforts to address the eventual transition from rhetoric to literary studies in the late-nineteenth-century university.

32. *Ibid.*, 14.

33. *Ibid.*, 10–11, emphasis added.

34. Quoted in Smith, "The 'American Creed,' " 233–34.

35. Jacobson, *Whiteness of a Different Color*, 42.

36. Jacobson, *Ibid.*, 8.

37. Jacobson, *Ibid.*, 68.

38. Smith, *Civic Ideals*, 228.

39. Although I am proposing a parallel between the rise of literary formalism and organicism in English studies and the rise of racial science and politics at the turn of the century, I do not want to suggest that a concern for aesthetics is inherently racist or that its late-nineteenth-century deployment overdetermines and cancels out the contemporary study or use of aesthetics for politically progressive ends. At the same time, I am interested in the connection between certain forms of literary discourse and nationalist extremism, or what David Carroll provocatively names "literary fascism." Carroll has effectively demonstrated how certain forms of extreme nationalism in France came to be formulated in literary terms by addressing "the totalizing tendencies implicit in literature itself and [how it] constitutes a technique or mode of fabrication, a form of fictionalizing or aestheticizing not just of literature but politics as well, and the transformation of the disparate elements of each into organic, totalized works of art" (7), He further suggests that such nationalist extremism is a logical extension of a number of fundamental aesthetic concepts, such as "the notion of the integrity of 'Man' as a founding cultural principle and political goal; of the totalized, organic unity of the artwork as both an aesthetic and political ideal; and finally, of culture considered as the model for the positive form of political totalization, the ultimate foundation for and the full realization and unification of both the individual and the collectivity" (7). See Carroll's *French Literary Fascism: Nationalism, Anti-Semitism, and the Ideology of Culture* (Princeton, N.J.: Princeton University Press, 1995).

40. See James A. Berlin's *Rhetoric, Poetics, and Cultures: Refiguring College English Studies* (Urbana, IL: NCTE, 1996); Franklin Court's *Institutionalizing English Literature. The Culture and Politics of Literary Study, 1750–1900* (Stanford: Stanford University Press, 1992); Sharon Crowley's *Composition in the University* (Pittsburgh: University of Pennsylvania Press, 1998); Terry Eagleton's *Literary Theory: An Introduction* (Minneapolis: Uuniversity of Minneapolis Press, 1983); Robert Scholes's *The Rise and Fall of English: Reconstructing English as a Discipline* (New Haven: Yale University Press, 1998); and Raymond Williams's, *Marxism and Literature* (Oxford: Oxford

University Press, 1977) for outstanding critical analyses of the rise of English studies in Britain and the United States.

41. Of course, counter-narratives of race (associated with the work of Franz Boas) and civic education (associated with John Dewey) emerged simultaneously with biological racism and the social efficiency in education. While Dewey remained in some ways marginal in spite of his significance to American educational philosophy, Boas gained greater recognition in the Thirties. See Matthew Frye Jacobson's, *Whiteness of a Different Color* for a commentary on Boas's challenge to biological racism. For an excellent analysis of Dewey's legacy see Herbert Kleibbard's, *The Struggle for the American Curriculum 1893–1958* (New York: Routledge, 1995).

42. Gerald Graff, *Professing Literature* (Chicago: University of Chicago Press, 1987), 20–28.

43. S. Michael Halloran, "Rhetoric in the American College Curriculum: The Decline in Public Discourse," *Pre/Text* 3: 3 (1982), 252.

44. Halloran, *Ibid.*, 246.

45. Julian P. Boyd, *The Papers of Thomas Jefferson, vol. 2, 1777 to 18 June 1779*, (Princeton, N.J.: Princeton University Press, 1950), 526–67. Emphasis added.

46. Quoted in Roy J. Honeywell, *The Educational Work of Thomas Jefferson* (Cambridge, Mass.: Harvard University Press, 1931), 250.

47. For an eloquent elaboration on the distinction between education and training, see Stanley Aronowitz's, *The Knowledge Factory* and Henry Giroux's, *Corporate Culture and the Attack on Higher Education and Public Schooling* (Bloomington, I.N.: Phi Delta Kappa Educational Foundation, 1999).

48. Boyd, *The Papers of Thomas Jefferson*, 526–27.

49. Quoted in Honeywell, *The Educational Work of Thomas Jefferson*, 148.

50. Merrill D. Peterson, *The Portable Thomas Jefferson* (New York: Penguin, 1975), 382.

51. Quoted in Honeywell, *The Educational Work of Thomas Jefferson*, 251.

52. Quoted in Honeywell, 251.

53. Of course, Native Americans didn't get the vote until 1924, 1926, or 1928 (depending on the interpretation of the law).

54. Peterson, *The Portable Thomas Jefferson*, 188.

55. Quoted in Steven E. Tozer, Paul C. Violas, and Guy Senese, *School and Society: Historical and Contemporary Perspectives*, 3rd ed. (Boston: McGraw Hill, 1998), 38.

56. Cremin, *American Education*, 390.

57. Of course, other factors played a role in the general educational movement away from the traditional classical curriculum. Though I won't elaborate on them here, I would argue they were no less racially motivated. Advances in modern psychology by figures like G. Stanley Hall, who advanced the general scientific proposition that "ontology recapitulates phylogeny" as a curriculum theory, and Edward L. Thorndike, who became the great apostle of the intelligence testing movement and its drive to place students in "inferior" and "superior" categories, also had an impact on the shift to a differentiated school curricula. See Kliebbard's, *The Struggle Over the American Curriculum* and Tozer's, *School and Society* for further elaboration on such contributions to "progressive" education.

58. Quoted in Tozer, *School and Society*, 110.

59. Alfred Schultz, *Race or Mongrel* (Boston: L.C. Page, 1908), 81.
60. Quoted in David Bennett, *The Party of Fear: The American Far Right from Nativism to the Militia Movement*, 2nd ed. (New York: Vintage, 1995), 221.
61. Calvin Coolidge, "Whose Country is This?" *Good Housekeeping* (February 1921), 14.
62. *Ibid.*
63. Hannaford, *Race*, 276.
64. *Ibid.*
65. Calvin Coolidge, *America's Need for Education* (Boston: Houghton, 1925), 51–52.
66. *Ibid.*, 13.
67. *Ibid.*, 38.
68. *Ibid.*, 47.
69. *Ibid.*, 44–45.
70. *Ibid.*, 49.
71. David Shumway, *Creating American Civilization: A Genealogy of American Literature as an Academic Discipline* (Minneapolis: University of Minneapolis Press, 1986), 19.
72. *Ibid.*, 38.
73. See Henry A. Giroux, *Schooling and the Struggle for Public Life* (Minneapolis: University of Minneapolis Press, 1988) for an analysis of the relationship between the institutional arrangement of schooling, critical pedagogical practices, and democratic public life.
74. Jeffrey Hart, "How to Get a College Education," *The New Republic* (September 1996), 38.
75. *Ibid.*
76. Ellen Willis, "We Need a Radical Left," *The Nation* (June 29, 1998), 18.
77. Stanley Fish, *Professional Correctness: Literary Studies and Political Change* (New York: Clarendon, 1995), 16.
78. Richard Rorty, "The Inspirational Value of Great Works of Literature," *Raritan* (Summer 1995), 15.
79. *Ibid.*, 13.
80. Lynn Hunt, "Democratization and Decline? The Consequences of Demographic Change in the Humanities," in *What Happened to the Humanities?*, ed. Alvin Kernan (Princeton, N.J.: Princeton University Press, 1997), 28.
81. Edward Said, *Representations of the Intellectual* (New York: Pantheon, 1994), 77.
82. *Ibid.*
83. Julie Drew, "Cultural Composition: Stuart Hall on Ethnicity and the Discursive Turn," in Gary A. Olson and Lynn Worsham, eds., *Race, Rhetoric and the Postcolonial* (Albany: State University of New York Press, 1999), 226.
84. Aronowitz, *The Knowledge Factory*, 169.
85. Gary A. Olson and Lynn Worsham, "Staging the Politics of Difference: Homi Bhabha's Critical Literacy," in *Race, Rhetoric and the Postcolonial*, 12.
86. Paulo Freire, *Pedagogy of Freedom: Ethics, Democracy, and Civic Courage* (Lanham, Md.: Rowman and Littlefield, 1998), 94.
87. See, for example, Harold Bloom, *The Western Canon* (New York: Harcourt Brace, 1994); Richard Rorty, "The Inspirational Value of Great Works of Literature," 8–17.
88. Hart, "How to Get a College Education," 38.
89. *Ibid.*

90. Judith Shklar, *American Citizenship*, 11.
91. Samuel Huntington, *American Politics: The Promise of Disharmony* (Cambridge, Mass.: Belknap, 1981), 23.
92. E. D. Hirsch, Jr., *Cultural Literacy: What Every American Needs to Know* (Boston: Houghton Mifflin, 1987), 22.
93. Stanley Aronowitz and Henry A. Giroux, *Postmodern Education: Politics, Culture and Social Criticism* (Minneapolis: University of Minneapolis Press, 1991), 48–49.
94. Graff, *Professing Literature*, 133–36. See also Lawrence W. Levine's, *The Opening of the American Mind: Canons, Culture, and History* (Boston: Beacon, 1996) for an incisive history of the "Western civ" debate.
95. Bloom, *The Western Canon*, 31.
96. Roger Shattuck, *Candor and Perversion: Literature, Education, and the Arts* (New York: Norton, 1999), 11.
97. *Ibid.*, 25.
98. *Ibid.*
99. *Ibid.*
100. *Ibid.*, 7.
101. *Ibid.*
102. *Ibid.*
103. *Ibid.* Emphasis added.
104. Freire, *Pedagogy of Freedom*, 90–91.

Chapter 5   The Return of the Ivory Tower: Black Educational Exclusion in the Post-Civil Rights Era

1. W. E. B. Du Bois, *Black Reconstruction in America: 1860–1880* (New York: Atheneum, 1992 [1935]), 377.
2. *Ibid.*, 346.
3. *Ibid.*, 634–35.
4. *Ibid.*, 714.
5. *Ibid.*, 724.
6. Claude Lemert, *Dark Thoughts: Race and the Eclipse of Society* (New York: Routledge, 2002), 223.
7. We would like to thank David Goldberg for reminding us just how far the rolling back of civil rights victories has gone. For an analysis of the limits of electoral reform, see Manning Marable's *The Great Wells of Democracy: The Meaning of Race in American Life* (New York: BasicCivitas Books, 2002). For an analysis of the role felony disenfranchisement played in the theft of the 2000 U.S. presidential election, see Greg Palast's *The Best Democracy Money Can Buy: The Truth About Corporate Cons, Globalization, and High-Finance Fraudsters* (New York: Pluto Press, 2002).
8. In place of rational argumentation, the public is bombarded with the cartoon-ish rant of right-wing extremists such as Ann Coulter, Michael Savage, Bill O'Reilly, Dinesh D'Souza, and Rush Limbaugh, at the same time that conservatives bemoan the "liberal" media.
9. William Greider, "Rolling Back the Twentieth Century," *The Nation* (May 12, 2003), 11–19.

10. Kevin Baker, "We're in the Army Now: The G.O.P.'s Plan to Militarize Our Culture," *Harper's Magazine* (October 27, 2003), 38–39.

11. Bill Moyers, "This is Your Story: The Progressive Story of America. Pass it On." Text of speech to the *Take Back America* Conference sponsored by the Campaign for America's Future, June 4, 2003. Washington, D.C. Available on line at: http://www.commondreams.org.

12. Lewis Lapham, "Yankee Doodle Dandy," *Harpers* (August 2003), 10.

13. *Ibid.*, 10.

14. See Bruce D. Dixon, "Muzzling the African American Agenda with Black Help: The DLC's Corporate Dollars of Destruction" in *Black Commentator,* available at: http://www.commondreams.org.

15. *Ibid.*

16. Howard Zinn, *A People's History of the United States: 1492–Present*, Revised and Expanded Edition (New York: Harper Perennial, 1995), 247.

17. Dalton Conley, *Being Black, Living in the Red* (Berkeley: University of California Press, 1999), 25.

18. See, for example, Manning Marable's *The Great Wells of Democracy: The Meaning of Race in Civic Life* (New York: BasicCivitas, 2002) and Howard Winant's *The World is a Ghetto: Race and Democracy after World War II* (New York: Basic, 2002).

19. Louis Uchitelle, "Blacks Lose Better Job Faster as Middle-Class Work Drops," *New York Times* (July 12, 2003), C14.

20. *Ibid.*, 20.

21. See "June Jobless Rate Among America's Teens Highest in 55 Years," press release by the Children's Defense Fund, available at: http://www. childrensdefense.org.

22. Mark Gongloff, "U.S., Jobs Jumping Ship," *CNNMoney* (July 22, 2003). Available on line at: http://cnnmoney.printthis.clickability.com.

23. Du Bois, *Black Reconstruction*, 678.

24. *Ibid.*, 674.

25. Loïc Wacquant, "From Slavery to Mass Incarceration: Rethinking the 'Race Question' in the U.S.," *New Left Review* 13 (January/February 2002), 56.

26. For an elaborated analysis of the racially coded rhetoric of the right, see Thomas Edsall and Mary Edsall, *Chain Reaction: The Impact of Race, Rights and Taxes on American Politics* (New York: Norton, 1992).

27. John Brenkman, "Race Publics," *Transition*, 1995, 17. Emphasis added.

28. *Ibid.*, 16.

29. Wacquant, 53. For further analysis of the linkages between slavery, convict leasing, and present-day mass incarceration, see Angela Y. Davis, "From the Convict Lease System to the Super Max" in Joy James, ed., *States of Confinement* (New York: St. Martin's Press, 2000), and Davis's "Racialized Punishment and Prison Abolition" in Joy James, ed., *The Angela Davis Reader* (Malden, Mass.: Blackwell, 1998), 96–107.

30. Quoted in: Jamal, Munia Abu, "Prisoners of war," *Monthly Review* (July–August 2001), 56.

31. Paul Street, "Race, Prisons, and Poverty: The Race to Incorporate in the Age of Correctional Keynesianism," *Z Magazine* (May 2001), 25–31.

32. Mark Mauer, *Race to Incarcerate* (New York: The New Press, 1999), 19.

33. James Vicini, "U.S. Prison and Jail Population Increases in 2002," *Reuters* (July 27, 2003). Available on line at: http://www.commondreams.org.

34. Graham Boyd, "The Drug War is the New Jim Crow," published by the NACLA Report on the Americas, July/August 2001. Available on line at: http://www.aclu.org/issues/drugpolicy/NACLA_Article.html.

35. Du Bois, *Black Reconstruction*, 713.

36. *Ibid.*, 712.

37. *Ibid.*, 717.

38. Charles Murray, *Losing Ground: American Social Policy 1950–1980* (New York: Basic, 1984), 443–44.

39. David Theo Goldberg, *The Racial State* (Malden, Mass.: Blackwell, 2002).

40. *Ibid.*

41. Du Bois, *Black Reconstruction*, 123.

42. *Ibid.*, 697.

43. *Ibid.*, 667.

44. Tamar Lewin and Jennifer Medina, "To Cut Failure Rate, Schools Shed Students," *The New York Times* (July 31, 2003), A1.

45. Gary Orfield, Susan E. Eaton and *The Harvard Project on School Desegregation. Dismantling Desegregation: The Quiet Reversal of Brown v. Board of Education* (New York: The New Press, 1996), 341.

46. *Ibid.*, 23.

47. Jonathan Kozol, *Savage Inequalities: Children in America's Schools* (New York: Harper Perennial, 1992), 55.

48. Orfield, 339.

49. *Ibid.*, 339.

50. See Henry Giroux, *The Abandoned Generation: Democracy Beyond the Culture of Fear* (New York: Palgrave, 2003); William Ayers et al., *Zero Tolerance: Resisting the Drive for Punishment in Our Schools* (New York: New Press, 2001); and Gary Orfield and Mindy Kornhaber, *Raising Standards or Raising Barriers? Inequality and High-Stakes Testing in Public Education* (New York: Century Foundation, 2001).

51. Paul Krugman, "State of Decline," *The New York Times* (August 1, 2003), A21.

52. Mike Davis, "Cry California," posted September 3, 2003. Available on line at: http://www.commondreams.org.

53. Senator Robert Byrd, "From Bad to Worse . . . Billions for War on Iraq, A Fraction for Poor Kids," Senate floor remarks, September 5, 2003. Available on line at: http://www.commondreams.org.

54. See the July 18, 2003 issue of the *Chronicle of Higher Education*, particularly the article by Stephen Burd, "Bush's Next Target?" A18–A20.

55. Greg Winter, "Tens of Thousands Will Lose College Aid, Report Says," *New York Times* (July 18, 2003), A12.

56. Quoted in Nat Hentoff, "Sandra Day O'Conner's Elitist Decision: Another Three Card Monte Game," *Village Voice* (July 18, 2003).

57. Quoted in Lani Guinier, "Saving Affirmative Action and a Process for Elites to Choose Elites," *Village Voice* (July 8, 2003), 1–5.

58. Burd, A20.

59. Cited in Stanley Fish, "Colleges Caught in a Vise," *The New York Times* (September 18, 2003), A31.

60. *Ibid.*
61. Cited in Janelle Brown, "Smoke a Joint and Your Future is McDonalds," *Salon* (May 20, 2002), 2. Available at: http://www.alternet.org/story.html? StoryID= 13168.
62. *Ibid*, 4.
63. Du Bois, *Black Reconstruction*, 708.
64. *Ibid.*, 678.
65. We would like to thank John Camoroff for bringing this point home in a personal exchange in July 2003.
66. Du Bois, *Black Reconstruction*, 714.
67. For a brilliant analysis of managing diversity, see David Theo Golberg, *The Racial State* (Malden, Mass.: Blackwell, 2002). Also, Manning Marable, ed. in *Dispatches from the Ebony Tower* (New York: Columbia University Press, 2001), and Henry Giroux, in *Impure Acts: The Practical Politics of Cultural Studies* (New York: Routledge, 2000).
68. Brenkman, "Race Publics," 14.
69. *Ibid.*, 14.
70. Jerry L. Martin and Anne O. Neal, "Defending Civilization: How Our Universities are Failing America and What Can be Done About It." Available on line at: www.goacta.org.
71. Carl Boggs, *The End of Politics. Corporate Culture and the Decline of the Public Sphere* (New York: Guilford Press), 135.
72. *Ibid.*, 137.
73. *Ibid.*
74. *Ibid.*
75. Kellogg, *The Chronicle of Higher Education*. January 2001, A47.
76. *Ibid.*
77. Gerald Early, "The Way Out of Here," *New York Times Book Review* (March 3, 2002), 12.

Chapter 6 Youth, Higher Education, and the Breaking of the Social Contract: Toward the Possibility of a Democratic Future

1. Ngugi Wa Thiong'o, *Moving the Centre: The Struggle for Cultural Freedoms* (Portsmouth, N.H.: Heinemann, 1993), 76.
2. For some excellent critical commentaries on various aspects of neoliberalism and its consequences, see Noam Chomsky, *Profit Over People: Neoliberalism and the Global Order* (New York: Seven Stories Press, 1999); Pierre Bourdieu, *Acts of Resistance: Against the Tyranny of the Market* (New York: The New Press, 1998); Pierre Bourdieu et al., *The Weight of the World: Social Suffering in Contemporary Society* (Stanford, Calif.: Stanford University Press, 1999); Robert W. McChesney, *Rich Media, Poor Democracy: Communication Politics in Dubious Times* (New York: The New Press, 1999); Zygmunt Bauman, *Work, Consumerism, and the New Poor* (Philadelphia: Open University Press, 1998); Zygmunt Bauman, *Society Under Siege* (Cambridge, England: Polity Press, 2002).

3. Lawrence Grossberg, "Why Does Neo-Liberalism Hate Kids? The War on Youth and the Culture of Politics," *The Review of Education/Pedagogy/Cultural Studies* 23: 2 (2001), 133.

4. *Ibid.*, 133.

5. Heather Wokusch, "Leaving Our Children Behind," *Common Dreams News Center* (July 8, 2002), available on line at www.commondreams.org.

6. These figures are taken from 2002 data released in 2003 by the U.S. Census Bureau.

7. These figures are taken from Child Research Briefs, "Poverty, Welfare, and Children: A Summary of the Data." Available on line at www.childtrends.org.

8. These figures are taken from Childhood Poverty Research Brief 2, "Child Poverty in the States: Levels and Trends From 1979 to 1998." Available on line at http://www.nccp.org.

9. These figures largely come from Children's Defense Fund, *The State of Children in America's Union: A 2002 Action Guide to Leave No Child Behind* (Washington, D.C.: Children's Defense Fund Publication, 2002), iv–v, 13.

10. Jennifer Egan, "To Be Young and Homeless," *The New York Times Magazine*, March 24, 2002, 35.

11. Wokusch, 1.

12. At the time this book was about to go to press, President Bush requested $87 billion additional funding for the Iraqi occupation, which was approved by Congress.

13. These figures are taken from Robert Kuttner, "War Distracts from Bush's Budget Cuts," *Common Dreams*. Available on line at www.commondreams.org.

14. Bauman, *Work, Consumerism, and the New Poor*, 77.

15. Noreena Hertz, *The Silent Takeover: Global Capitalism and the Death of Democracy* (New York: The Free Press, 2001), 11.

16. Pierre Bourdieu, *Acts of Resistance*; Pierre Bourdieu et al., *The Weight of the World*.

17. Bauman, *In Search of Politics*, 2.

18. Naomi Klein, *No Logo* (New York: Picador, 1999), 177.

19. Grossberg, "Why Does Neo-Liberalism Hate Kids?" 133.

20. Cited in Frank Hearn, *Reason and Freedom in Sociological Thought* (Boston: Unwin Hyman, 1985), 175. The classic statements by Dewey on this subject can be found in John Dewey, *Democracy and Education* (New York: The Free Press, 1997: reprint edition—1st ed., 1916); See also, John Dewey, *The Public and Its Problems* (Columbus: Ohio University Press, 1954).

21. C. Wright Mills, *Power, Politics, and People*, ed. Irving Louis Horowitz (New York: Oxford University Press, 1963).

22. Jean Comaroff and John L. Comaroff, "Millennial Capitalism: First Thoughts on a Second Coming," *Public Culture* 12, no. 2 (Duke University Press, 2000), 305–06.

23. Peter Beilharz, *Zygmunt Bauman: Dialectic of Modernity* (London: Sage, 2000), 160.

24. Comaroff and Comaroff, 332.

25. James Traub, "This Campus is Being Simulated," *The New York Times Magazine*, November 19, 2000, 93.

26. Cited in Roger Simon, "The University: A Place to Think?" in Henry A. Giroux and Kostas Myrsiades, eds., *Beyond the Corporate University* (Lanham, Md.: Rowman and Littlefield, 2001), 47–8.

27. Larry Hanley, "Conference Roundtable," *Found Object* 10 (Spring 2001), 103.
28. Masao Miyoshi, " 'Globalization,' Culture, and the University," in Fredric Jameson and Masao Miyoshi, eds., *The Cultures of Globalization* (Durham, N.C.: Duke University Press, 1998), 263.
29. Beilharz, 161.
30. Jacques Derrida, "Intellectual Courage: An Interview," *Culture Machine,* vol. 2 (2000), 9.
31. Michael Hanchard, "Afro-Modernity: Temporality, Politics, and the African Diaspora," *Public Culture* 11: 1 (Winter 1999), 253.
32. *Ibid.,* 256.
33. Jerome Bind, "Toward an Ethic of the Future," *Public Culture* 12: 1 (2000), 52.
34. Cornelius Castoriadis, "The Problem of Democracy Today," *Democracy and Nature* 8 (April 1996), 24.
35. Roger I. Simon, "On Public Time," Ontario Institute for Studies in Education, unpublished paper (April 1, 2002), 4.
36. Simon Critchley, "Ethics, Politics, and Radical Democracy—The History of a Disagreement," *Culture Machine,* available at www.culturemachine.tees.ac.uk.
37. James Rule, "Markets In their Place," *Dissent* (Winter 1998), 30.
38. Peter Euben, "Reforming the Liberal Arts," *The Civic Arts Review* 2 (Summer–Fall 2000), 8.
39. Cary Nelson, "Between Anonymity and Celebrity: The Zero Degrees of Professional Identity," *College English* 64: 6 (July 2002), 717.
40. Comaroff and Comaroff, 306.
41. Geoff Sharp, "The Idea of the Intellectual and After," in Simon Cooper, John Hinkson, and Geoff Sharp, eds., *Scholars and Entrepreneurs: The University in Crisis* (Melbourne, Australia: Arena Publications, 2002), 280.
42. Ben Agger, "Sociological Writing in the Wake of Postmodernism," *Cultural Studies/Cultural Methodologies* 2: 4 (November 4, 2002), 444.
43. Gary Rhoades, "Corporate, Techno Challenges, and Academic Space," *Found Object* 10 (Spring 2001), 143.
44. Stanley Aronowitz, "The New Corporate University," *Dollars and Sense* (March/April 1998), 32.
45. *Ibid.,* 444.
46. Rhoades, 122.
47. *Ibid.*
48. Nelson, 713.
49. Taken from the title of James Howard Kunstler's, *The Geography of Nowhere* (New York: Touchstone, 1993).
50. The most extensive analysis of the branding of culture by corporations can be found in Naomi Klein, *No Logo.*
51. Jeffrey L. Williams, "Franchising the University," in Henry A. Giroux and Kostas Myrsiades, eds., *Beyond the Corporate University: Culture and Pedagogy in the New Millennium* (Lanham, Md.: Rowman and Littlefield, 2001), 23.
52. Michael Peters, "The University in the Knowledge Economy," in Simon Cooper, John Hinkson, and Geoff Sharp, eds., *Scholars and Entrepreneurs: The University in Crisis* (Melbourne, Australia: Arena Publications, 2002), 148.
53. Zygmunt Bauman, *Globalization: The Human Consequence* (New York: Columbia University Press, 1998), 81.

54. Geoff Sharp, "The Idea of the Intellectual and After," in Cooper, Hinkson, and Sharp, eds., *Scholars and Entrepreneurs*, 275.

55. See John Hinkson, "Perspectives on the Crisis of the University," in Cooper, Hinkson, and Sharp, eds., *Scholars and Entrepreneurs*, 233–67.

56. *Ibid.*, 284–85.

57. Nick Couldry, "Dialogue in an Age of Enclosure: Exploring the Values of Cultural Studies," *The Review of Education/Pedagogy/Cultural Studies* 23: 1 (2001), 17.

58. Hinkson, "Perspectives on the Crisis of the University," in Cooper, Hinkson, and Sharp, eds., *Scholars and Entrepreneurs*, 233–67.

59. *Ibid.*, 259.

60. Arundhati Roy, *Power Politics* (Cambridge, Mass.: South End Press, 2001), 6.

61. The ideas on public intellectuals are taken directly from Edward Said, *Reflections on Exile and Other Essays* (Cambridge, Mass.: Harvard University Press, 2001), 502–03. For the reference to realist utopias, see Pierre Bourdieu, "For a Scholarship with Commitment," *Profession* (2000), 42.

62. Toni Morrison, "How Can Values Be Taught in the University?," *Michigan Quarterly Review* (Spring 2001), 278.

63. Cornelius Castoriadis, "Culture in a Democratic Society," in *The Castoriadis Reader*, ed. David Ames Curtis (Malden, Mass.: Blackwell, 1997), 343.

64. Cornelius Castoriadis, "The Nature and Value of Equity," in *Philosophy, Politics, Autonomy: Essays in Political Philosophy* (New York: Oxford University Press, 1991), 140.

65. Cornelius Castoriadis, "The Problem of Democracy Today," *Democracy and Nature*, vol. 8 (April 1996), 24.

66. Cornelius Castoriadis, "The Greek Polis and the Creation of Democracy," in *Philosophy, Politics, Autonomy*, 112.

67. Roy, 3.

68. Christopher Newfield, "Democratic Passions: Reconstructing Individual Agency," in Russ Castronovo and Dana Nelson, eds., *Materializing Democracy* (Durham, N.C.: Duke University Press, 2002), 314.

69. Ruth Levitas, "The Future of Thinking About the Future," in Jon Bird et al., *Mapping the Futures* (New York: Routledge, 1993), 257.

70. Cited in Bauman, *Work, Consumerism and the New Poor*, 98.

71. Jacques Derrida, "The Future of the Profession or the Unconditional University," in *Derrida Downunder*, Laurence Simmons and Heather Worth, eds. (Auckland, New Zealand: Dunmore Press, 2001), 7.

72. Samin Amin has captured this sentiment in his comment: "Neither modernity nor democracy has reached the end of its potential development. That is why we prefer the term 'democratization,' which stresses the dynamic aspect of a still-unfinished process, to the term 'democracy,' which reinforces the illusion that we can give a definitive formula for it." See Samir Amin, "Imperialization and Globalization," *Monthly Review* (June 2001), 12.

73. Alain Badiou, *Ethics: An Essay on the Understanding of Evil* (London: Verso, 2001), 96.

74. Ron Aronson, "Hope After Hope?" *Social Research* 66: 2 (Summer 1999), 489.

75. Roger Simon, "Empowerment as a Pedagogy of Possibility," *Language Arts* 64: 4 (April 1987), 371.

76. Cited in Gary A. Olson and Lynn Worsham, "Changing the Subject: Judith Butler's Politics of Radical Resignification," *JAC* 20: 4 (2000), 765.

77. Zygmunt Bauman and Keith Tester, *Conversations with Zygmunt Bauman* (Malden, Mass.: Polity Press, 2001), 4.

78. Lawrence Grossberg, "Introduction: Bringing It All Back Home—Pedagogy and Cultural Studies," in Henry A. Giroux and Peter McLaren, eds., *Between Borders: Pedagogy and the Politics of Cultural Studies* (New York: Routledge, 1994), 14.

79. Bauman and Tester, 20.

80. Jeffrey C. Goldfarb, "Anti-Ideology: Education and Politics as Democratic Practices," in Castronovo and Nelson, 345–67.

81. Stanley Aronowitz, "Introduction," in Paulo Freire, *Pedagogy of Freedom* (Lanham, Md.: Rowman and Littlefield, 1998), 10–11.

82. Cornelius Castoriadis, "Democracy as Procedure and Democracy as Regime," *Constellations* 4: 1 (1997), 4–5.

83. Chandra Talpade Mohanty, "On Race and Voice: Challenges for Liberal Education in the 1990s," *Cultural Critique* 14 (Winter 1989–1990), 192.

84. Cited in Paul Berman, "The Philosopher-King is Mortal," *The New York Times Magazine* (May 11, 1997), 36.

Chapter 7   Neoliberalism Goes to College: Higher Education in the New Economy

1. This quote is taken from George Gilder's "Recapturing the Spirit of Enterprise" and is cited on line at http://www.geocities.com.

2. Zygmunt Bauman, *Society Under Siege* (Malden, Mass.: Blackwell, 2002), 68.

3. Susan Buck-Morss, *Thinking Past Terror: Islamism and Critical Theory on the Left* (New York/London: Verso, 2003), 29.

4. These ideas are taken from Benjamin R. Barber's "Blood Brothers, Consumers, or Citizens? Three Models of Identity—Ethnic, Commercial, and Civic," in Carol Gould and Pasquale Pasquino, eds., *Cultural Identity and the Nation State* (Lanham, Md.: Rowman and Littlefield, 2001), 65.

5. Noreena Hertz, *The Silent Takeover: Global Capitalism and the Death of Democracy* (New York: Free Press, 2001), 66.

6. Zygmunt Bauman, *In Search of Politics* (Stanford, Calif.: Stanford University Press, 1999), 127.

7. The classic texts defending corporate culture include: Terrance Deal and Alan Kennedy, *Corporate Culture: The Rites and Rituals of Corporate Life* (Reading, Mass.: Addison-Wesley, 1982) and Thomas Peterson and Robert Waterman, *In Search of Excellence* (New York: Harper and Row, 1982). I also want to point out that corporate culture is a dynamic, ever-changing force. But in spite of its innovations and changes, it rarely if ever challenges the centrality of the profit motive, or fails to prioritize commercial considerations over a set of values that would call the class-based system of capitalism into question. For an informative discussion of the changing nature of corporate culture in light of the cultural revolution of the 1960s, see Thomas Frank, *The Conquest of Cool* (Chicago: University of Chicago Press, 1997).

8. Alan Bryman, *Disney and His Worlds* (New York: Routledge, 1995), 154.

9. Robin D. G. Kelley, *Yo' Mama's Disfunktional: Fighting the Culture Wars in Urban America* (Boston: Beacon Press, 1997).

10. Cited in Molly Ivins, "Bush Discovers Hunger and Looks the Other Way," *Common Dreams* (December 17, 2002), 1.

11. For a context from which to judge the effects of such cuts on the poor and children of America, see Children's Defense Fund, *The State of Children in America's Union: A 2002 Action Guide to Leave No Child Behind* (Washington, D.C.: Children's Defense Fund, 2002). For a critical and excellent commentary on Bush's tax cuts, see Paul Krugman, *Fuzzy Math* (New York: Norton, 2001). On the emergence of the prison-industrial complex and how it diverts money from higher education, see Henry A. Giroux, *Public Spaces, Private Lives: Democracy Beyond 9/11* (Lanham, Md.: Rowman and Littlefield, 2003); Christian Parenti, *Lockdown America: Police and Prisons in the Age of Crisis* (London: Verso Press, 1999); March Mauer and Meda Chesney-Lind, eds., *Invisible Punishment: The Collateral Consequences of Mass Imprisonment* (New York: The New Press, 2002).

12. Cited in Rock the Vote education page, which is available on line: www.rockthevote.org/issues_education.html.

13. Paul Krugman, "AARP Gone Astray," *New York Times* (November 21, 2003), A31.

14. This example is taken from Colin Leys, *Market-Driven Politics* (London: Verso, 2001), 212–13.

15. Pierre Bourdieu, "The Essence of Neoliberalism," *Le Monde Diplomatique* (December 1998). On line at http://www.en.monde-diplomatique.fr/1998/12/08bourdieu. p. 4.

16. Benjamin R. Barber, "A Failure of Democracy, Not Capitalism," *The New York Times* (Monday, July 29, 2002), A23.

17. Cited in Tom Turnipseed, "Crime in the Suites Enabled by Political Corruption Causes a Crisis in the Credibility of U.S. Capitalism," *Common Dreams* (Saturday, June 29, 2002), 1. Available on line at www.commondreams.org.

18. For a recent defense of this tradition, see Anne Colby, Thomas Ehrlick, Elizabeth Baumont, and Jason Stephens, *Educating Citizens* (San Francisco: Jossey-Bass, 2003).

19. Sheila Slaughter, "Professional Values and the Allure of the Market," *Academe* (September–October, 2001), 1.

20. Cited in Beth Huber, "Homogenizing the Curriculum: Manufacturing the Standardized Student." Available on line at http://www.louisville.edu/journal/workplace/huber.html. p. 1.

21. John Palattella, "Ivory Towers in the Marketplace," *Dissent* (Summer 2001), 73.

22. See Ronald Strickland, "Gender, Class and the Humanities in the Corporate University," *Genders* 35 (2002), 2. Available on line at www.genders.org.

23. Patrick Healy, "College Rivalry," *The Boston Globe Magazine* (June 29, 2003). Available on line at http://www.boston.com/globe/magazine/2003/0629/coverstory entire.htm. p. 1.

24. Critical educators have provided a rich history of how both public and higher education have been shaped by the politics, ideologies, and images of industry. For example, see Samuel Bowles and Herbert Gintis, *Schooling in Capitalist America* (New York: Basic Books, 1976); Michael Apple, *Ideology and Curriculum* (New York: Routledge, 1977); Martin Carnoy and Henry Levin, *Schooling and Work in the Democratic State* (Stanford, Calif.: Stanford University Press, 1985); Stanley Aronowitz and Henry A. Giroux, *Education Still Under Siege* (Westport, Conn.: Bergin and Garvey, 1993); Stanley

Aronowitz and William DiFazio, *The Jobless Future* (Minneapolis: University of Minnesota Press, 1994); Cary Nelson, ed., *Will Teach for Food: Academic Labor in Crisis* (Minneapolis: University of Minnesota Press, 1997); D.W. Livingstone, *The Education-Jobs Gap* (Boulder, Col.: Westview, 1998).

25. See Robert Zemsky, "Have We Lost the 'Public' in Higher Education?" *The Chronicle of Higher Education* (May 30, 2003), B9.

26. See Robert Zemsky, "Have We Lost the 'Public' in Higher Education?" *The Chronicle of Higher Education* (May 30, 2003), B7-B9; Derek Bok, *Universities in the Marketplace: The Commercialization of Higher Education* (Princeton, N.J.: Princeton University Press, 2003).

27. Richard Ohmann, "Citizenship and Literacy Work: Thoughts Without a Conclusion," *Workplace* (November 25, 2002). Available on line: http://www.louisville.edu/journal/workplace/issue7/ohmann.html.

28. Bok, *Universities in the Marketplace*, 65.

29. *Ibid.*, 70.

30. Cited in Katherine S. Mangan, "Medical Schools Routinely Ignore Guidelines on Company-Sponsored Research, Study Finds," *The Chronicle of Higher Education* (October 25, 2002). Available on line: http://chronicle.com/daily/2002/10/2002102250ln.htm. p. 1.

31. Cited in Lila Guterman, "Conflict of Interest is Widespread in Biomedical Research, Study Finds," *The Chronicle of Higher Education* (January 22, 2003). Available on line: http://chronicle.com/daily/2003/01/2003012202n.htm. p. 1.

32. *Ibid.*

33. Cited in Katherine S. Mangan, 1.

34. Bok, *Universities in the Marketplace*, 75.

35. *Ibid.*, 200–01.

36. On this issue, see Bauman, *In Search of Politics*.

37. Stanley Aronowitz, "The New Corporate University," *Dollars and Sense* (March/April, 1998), 32–35. On cost of Iraq War, see Tom Shanker, "Rumsfeld Reveals $6bn a Month for War," *The Sydney Morning Herald* (July 11, 2003). Available on line: http://www.smh.com.au/articles/2003/07/10/1057783288802.html.

38. The many books extolling corporate CEOs as a model for leadership in any field is too extensive to cite, but one typical example can be found in Robert Heller, *Roads to Success: Put Into Practice the Best Business Ideas of Eight Leading Gurus* (New York: Dorling Kindersley, 2001).

39. For a devastating critique of corporate culture and corruption, see Arianna Huffington, *Pigs at the Trough* (New York: Crown Books, 2003), 14. William Greider provides another analytical register by arguing that the neoliberal and neoconservative right wing is trying to roll back the twentieth century to the period before the New Deal. See William Greider, "The Right's Grand Ambition: Rolling Back the 20th Century," *The Nation* (May 12, 2003), 11–12, 14, 16–19.

40. For an excellent analysis of Michael Milken's role in various education projects, see Robin Truth Goodman and Kenneth Saltman, *Strange Love: Or How We Learn to Stop Worrying and Love the Market* (Lanham, Md.: Rowman and Littlefield, 2002).

41. Catherine S. Manegold, "Study Says Schools Must Stress Academics," *The New York Times*, Friday (September 23, 1998), A22. It is difficult to understand how

any school system could have subjected students to such a crude lesson in commercial pedagogy.

42. David L. Kirp, "Education for Profit," *The Public Interest* 152 (Summer 2003), 105.

43. *Ibid.*, 112.

44. George Soros, *On Globalization* (New York: Public Affairs, 2002), 164, 6.

45. Stanley Aronowitz and William DiFazio, "The New Knowledge Work," in A. H. Halsey, Hugh Lauder, Phillip Brown, Amy Stuart Wells, eds., *Education: Culture, Economy, Society* (New York: Oxford, 1997), 193.

46. This issue is explored in great detail in Robert W. McChesney, *Rich Media, Poor Democracy: Communication Politic in Dubious Times* (New York: The New Press, 1999).

47. This is amply documented in Jeremy Rifkin, *The End of Work* (New York: G. Putnam Book, 1995); William Wolman and Anne Colamosca, *The Judas Economy: The Triumph of Capital and the Betrayal of Work* (Reading: Addison-Welsley Publishing, 1997); Stanley Aronowitz and William DiFazio, *The Jobless Future* (Minneapolis: University of Minnesota Press, 1994); *The New York Times Report: The Downsizing of America* (New York: Times Books, 1996); Stanley Aronowitz and Jonathan Cutler, *Post-Work* (New York: Routledge, 1998).

48. Cornel West, "The New Cultural Politics of Difference," *October* 53 (Summer 1990), 35.

49. Sheila Slaughter, "Professional Values and the Allure of the Market," *Academe* (September–October, 2001), 3–4.

50. Jeffrey Selingo, "Reform Plan or 'Corporate Takeover'?" *The Chronicle of Higher Education* (April 18, 2003), A28.

51. Cited in NEA Higher Education Research Center, *Update* 7: 3 (June 2001), 1.

52. Sharon Walsh, "Study Finds Significant Increase in Number of Part-Time and Non-Tenure-Track Professors," *The Chronicle of Higher Education* (October 29, 2002), B1. Available on line: http://chronicle.com/daily/2002/10/ 2002102904n. htm. The full text of the American Council of Education study can be found on the council's web site.

53. David F. Noble, "The Future of the Digital Diploma Mill," *Academe* 87: 5 (September–October 2001), 29. These arguments are spelled out in greater detail in David F. Noble, *Digital Diploma Mills: The Automation of Higher Education* (New York: Monthly Review Press, 2002).

54. Noble, "The Future of the Digital Diploma Mill," 31.

55. Both Levine's statement and the following quote can be found in Eyal Press and Jennifer Washburn, "Digital Diplomas," *Mother Jones* (January/February 2001), 2. Available on line at http://www.motherjones.com.

56. Press and Washburn, 2.

57. Herbert Marcuse, "Some Social Implications of Modern Technology," in *Technology, War, and Fascism*, ed. Douglas Kellner (New York: Routledge, 1998), 45.

58. Zygmunt Bauman, *Modernity and the Holocaust* (Ithaca, N.Y.: Cornell University Press, 1989).

59. *Ibid.*

60. For a critical analysis of the flaws and possibilities of such approaches in higher education, see David Trend, *Welcome to Cyberschool: Education at the*

Crossroads in the Information Age (Lanham, Md.: Rowman and Littlefield, 2001); Andrew Feenberg, *Questioning Technology* (New York: Routledge, 1999); Hubert L. Dreyfus, *On the Internet* (New York: Routledge, 2001); Mark Poster, *What is the Matter with the Internet?* (Minneapolis: University of Minnesota Press, 2001).

61. Noble, *Digital Diploma Mills.*

62. For more details on the creation of on line degrees for corporations, see Dan Carnevale, "Colleges Tailor On line Degrees for Individual Companies," *The Chronicle of Higher Education* (January 28, 2002). Available on line at http://chronicle.com/cgi2-bin/printable.cgi. p. 1.

63. Andrew Feenberg, *Questioning Technology*, viii.

64. *Ibid.*, xv.

65. Slaughter, 3.

66. For an extensive analysis of the issue of intellectual property rights and the control over academic work in the university, see Corynne McSherry, *Who Owns Academic Work?: Battling for Control of Intellectual Property* (Cambridge, Mass.: Harvard University Press, 2002).

67. Julia Porter Liebeskind, "Risky Business: Universities and Intellectual Property," *Academe* (September–October 2001). Available on line at www.aaup.org/publications/Academe/01SO/so01lie.htm. p. 2.

68. *Ibid.*, 2.

69. Bok, *Universities in the Marketplace*, 172.

70. Press and Washburn, "Digital Diplomas," 8.

71. Condren cited in *Ibid.*, 8.

72. Stanley Aronowitz, "The New Corporate University," *Dollars and Sense* (March/April 1998), 34–35.

73. This issue is taken up in Michael Berube, "Why Inefficiency is Good for Universities," *The Chronicle of Higher Education* (March 27, 1998), B4-B5.

74. Cited in Action Alert, *Rock the Vote Issues in Education* (July 12, 2003). Available on line: http:www.rockthevote.org.

75. Jeff Williams, "Brave New University," *College English* 61: 6 (July 1999), 740.

76. This information is taken from Editorial, "Pricing the Poor Out of College," *The New York Times* (March 27, 2002), A27.

77. Stephen Burd, "Lack of Aid Will Keep 170,000 Qualified, Needy Students Out of College This Year, Report Warns," *The Chronicle of Higher Education* (Thursday, June 27, 2002), 1. Available on line at: http://chronicle.com/daily/2002/06/2002062701n.html. For a robust argument for making college free for all students, see Adolph L. Reed, Jr., "Free College for All," *The Progressive* (April 2002), 12–14.

78. *Ibid.*

79. Michael Margolis, "Brave New Universities," *FirstMonday* (January 2002), 2. Available on line at http://www.firstmonday.dk/issues3_5/margolis/.

80. For a variety of critical commentaries on this issue, see Nelson, *Will Teach for Food*; Martin, *Chalk Lines*; Cary Nelson and Stephen Watt, *Academic Keywords: A Devil's Dictionary for Higher Education* (New York: Routledge, 1999); Aronowitz, *The Knowledge Factory.*

81. Cary Nelson, *Manifesto of a Tenured Radical* (New York: NYU Press, 1997), 169.

82. Trend, *Welcome to Cyberschool*, 55.

83. Jennifer L. Croissant, "Can This Campus be Bought?" *Academe* (September–October 2001). Available on line at www.aaup.org/publications/ Academe01SO/so01cro.html. p. 2.

84. Pierre Bourdieu, *Acts of Resistance* (New York: New Press, 1999), 8.

85. Peter Euben, "Reforming the Liberal Arts," *The Civic Arts Review*, no. 2 (Summer–Fall 2000), 10.

86. Bill Readings, *The University in Ruins* (Cambridge, Mass.: Harvard University Press, 1996). As Ronald Strickland points out, Readings has almost nothing to say about how the corporate university reproduces the academic and social division of labor between elite and second-tier universities that is so central to the changing global economic landscape. See Strickland, 1–16.

87. Williams, "Brave New University," 750.

88. Noam Chomsky, "Paths Taken, Tasks Ahead," *Profession* (2000), 35.

89. Pierre Bourdieu, "For a Scholarship of Commitment," *Profession* (2000), 45, 42–43.

90. "A Conversation between Lani Guinier and Anna Deavere Smith, 'Rethinking Power, Rethinking Theater,' " *Theater* 31: 1 (Winter 2002), 36.

91. Robin D. G. Kelley, "Neo-Cons of the Black Nation," *Black Renaissance Noire* 1: 2 (Summer/Fall 1997), 146.

92. See Nelson, *Will Teach For Food*.

93. Aronowitz, *The Knowledge Factory*, 63, 109–110.

94. "Students Hold 'Teach-Ins' to Protest Corporate Influence in Higher Education," *The Chronicle of Higher Education* (March 13, 1998), A11. Also note Peter Dreier's commentary on the student activism emerging in 1998, in "The Myth of Student Apathy," *The Nation* (April 13, 1998), 19–22.

95. Liza Featherstone, "Sweatshops, Students, and the Corporate University," *Croonenbergh's Fly* no. 2 (Summer 2002), 108.

96. For a commentary on students against sweatshops, see *Ibid.*, 107.

97. See, for example, Liza Featherstone, *Students Against Sweatshops* (London: Verso, 2002); Alexander Cockburn, Jeffrey St. Clair, and Allan Sekula, *5 Days That Shook the World: Seattle and Beyond* (London: Verso, 2000).

98. We have taken this idea from Nick Couldry, "A way out of the (televised) endgame?" *OpenDemocracy*, available on line at http://www.opendemocracy.net/forum/strands_home.asp. p. 1.

99. Imre Szeman, "Introduction: Learning to Learn from Seattle," in Imre Szeman, ed. (Special Double Issue) "Learning From Seattle." *The Review of Education, Pedagogy, and Cultural Studies* 24: 1–2 (January–June, 2002), 5.

100. *Ibid.*, 4.

101. *Ibid.*, 5.

102. Jacques Derrida, "Intellectual Courage: An Interview," *Culture Machine*, vol. 2 (2000), 9. Available at http://culturemachine.tees.ac.uk/articles/art_derr.htm.

103. Both quotes are from Eric Gould, *The University in a Corporate Culture*, 225.

104. These ideas are taken from Susan Buck-Morss, *Thinking Past Terror*.

105. Cited in "Why Does Neo-Liberalism Hate Kids? The War on Youth and the Culture of Politics," *The Review of Education/Pedagogy/Cultural Studies* 23: 2 (2001), 114.

Index